Risk and Technological Culture

T0186394

The question as to whether we are now entering a risk society has become a key debate in contemporary social theory. *Risk and Technological Culture* presents a critical discussion of the main theories of risk – from Ulrich Beck's foundational work to that of his contemporaries such as Anthony Giddens and Scott Lash – and assesses the extent to which risk has impacted on modern societies. In this discussion Van Loon demonstrates how new technologies are transforming the character of risk and examines the relationship between technological culture and society through substantive chapters on topics such as waste, emerging viruses, communication technologies and urban disorders. In so doing this innovative new book extends the debate to encompass theorists such as Bruno Latour, Donna Haraway, Gilles Deleuze, Felix Guattari and Jean-François Lyotard.

Joost Van Loon is Senior Lecturer in Social Theory at the Nottingham Trent University. He is co-editor of *The Risk Society and Beyond* (Sage, 2000) and *Trust and Co-operation* ('t Spinhuys, 2000), and co-founding editor of *Space and Culture*.

Risk and Technological Culture

Towards a sociology of virulence

Joost Van Loon

Routledge
Taylor & Francis Group
LONDON AND NEW YORK

First published 2002
by Routledge
2 Park Square, Milton Park, Abingdon, Oxon OX14 4RN

Simultaneously published in the USA and Canada
by Routledge
711 Third Avenue, New York, NY 10017

Routledge is an imprint of the Taylor & Francis Group, an informa business

© 2002 Joost Van Loon

Typeset in Goudy by Keyword Typesetting Services Ltd

British Library Cataloguing in Publication Data
A catalogue record for this book is available from the British Library

Library of Congress Cataloging in Publication Data
A catalogue record for this book is available from the Library of Congress

ISBN 0-415-22900-6 (hbk)
ISBN 0-415-22901-4 (pbk)

Contents

Preface

On 11 September 2001, exactly 28 years after the murder of the Chilean president Salvatore Allende by an alliance between a military junta led by General Pinochet and the Central Intelligence Agency (CIA), a new chapter was opened in the history of modernity. As it happened, I was putting the last finishing touches to this book, wondering how to introduce the general argument that risk engenders a turning in modern, western culture towards an apocalyptic sensibility. On that day, three hijacked planes were deliberately crashed into the most central nodal points of the 'new' globalized world order: the World Trade Center and the Pentagon. With a devastating and cold precision, these calculated and coordinated series of attacks on both civilian lives as well as symbols of global power ground the world economy to a temporary halt. At the same time, centres of control were not functioning because people were being evacuated as a result of a heightened sense of vulnerability. Chaos and pandemonium prevailed as media encircled these sites of devastation and had nothing to report but some very bare facts, which were subsequently engrossed in immense speculation.

Although terrorism does not feature in this book at all, the similarities between terrorism risks and other risks that are being discussed here are remarkable. The immediate response to the actual catastrophe revealed that the security concepts deployed in terms of anti-terrorism policing did not adequately cover the open-ended nature or virtuality of risks. For example, the way in which domestic (as distinct from international) air travel operates in the USA shows that either there was no sense of risk associated with using aeroplanes as missiles, or that such risks were unduly ignored because of other perceived risks such as the costs of impeding speed of movement. Moreover, the way in which security concepts were framed in military terms, with an emphasis on large-scale high-tech mass destruction apparatuses, has been exposed as wholly inadequate in the face of the strategies of 'terrorism' being deployed on that day.

Above all, it showed that risks are potent only in their 'becoming real'. Once the catastrophe is inaugurated, they have already moved to somewhere else, for example, in the form of speculations about further attacks (such as bio-terrorism) or consequences in terms of world peace or economic stability.

Many commentators went so far as to suggest that this was an attack on a whole civilization, thereby unduly polarizing the issue in terms of global politics. As in the Gulf War, television was a primary site at which this cyberwarfare took place. It provided the primary stage for the theatrical demonstration that the most powerful nation in the global order can be severely damaged, and it enabled an outpouring of emotive discursive practices that bypassed almost all forms of diplomacy, tact and reason. Within a few hours, the world had become a very dangerous place indeed. The lack of certainty was easily filled with speculation, especially about who might be responsible, the actual death toll and the USA's most likely response. Indeed, suddenly it seemed as if the end of the world is nigh.

In this book, however, the rise of apocalypse culture is seen as much less dramatic and spectacular. In fact, I argue that what has manifested itself so suddenly on 11 September 2001 was already rooting in the surface of modernity. This is not a prophetic or visionary statement. Instead, the signs were already written on the wall more than 200 years ago with the advent of modernity. Despite all the rhetoric about the centrality of reason and deliberation as marking the essence of enlightened modern civilization, far more primordial and 'barbaric' sentiments have – on more than a few occasions – easily resurfaced. The history of modernity can to a large extent be written in the language of violence, often mystified as natural forms of domination (e.g. capitalism, colonialism, patriarchy). Whereas we may be shocked by the events of 11 September, we should perhaps not be surprised.

In the face of such devastation, it is all too easy to think that everything else pales in significance. Yet, the US Army killed more people in the nuclear bombing of Hiroshima, and more Iraqis died during the Gulf War. Indeed, in terms of death toll, the attacks are nothing compared to the atrocities committed against civilians in Rwanda or even Bosnia and Kosovo. Yet, these deaths are less spectacular not only because they have taken place in the margins of the global order, but also because they have taken much longer to enter into 'reality'.

Compared to terrorism and warfare, the deaths caused by ecological catastrophes or epidemics have raised far less concern, yet – when taking a long-term global perspective – they are much higher in number. The fact that risks associated with these phenomena are much less spectacular and as a result downplayed during times of crisis signals that risks are intricately linked to processes of signification and valorization. At the same time, such unnoticed and unanticipated hazards may in the long term, however, have just as devastating consequences, if not more. This is why it is worthwhile to consider risk as an immensely important force in the unfolding of the history of modern society. Following specific risks, we will uncover some essential elements of the logic of modernity which have the uncanny characteristic of engendering a turning within modernity itself. Risks expose the madness of reason, the self-destructive propensity of modern thought, the inevitability that a reckoning will occur, at some stage. Maybe it has already happened.

Acknowledgements

This book would not have been possible without the immense support I received from countless people, many of whom I will undoubtedly have failed to mention and to whom I sincerely apologize. First of all, I would like to thank John Urry and Mari Shullow for helping me to conceptualize the book and its outline and for their continued support throughout the years it has taken to come into being. I also thank the anonymous reviewers for helping me to see more clearly the errors of my ways, and Nottingham Trent University for giving me additional support to do the research and writing. I am most grateful to the many dear friends and colleagues with whom I have tried out different ideas over the years: Rob Shields, Ian Roderick, Neil Turnbull, Irene Hardill, Francesca Bargiela, Deborah Chambers, Harry Wels, Ulrich Beck, Willem Koot, Neal Curtis, Mike Featherstone, John Tomlinson, Marsha Smith, Stuart Allan, Ian Welsh, Cindy Carter, Rosemary McKechnie, Barbara Adam, Sarah Franklin, Claudia Castaneda, Bruno Latour, Marc Berg, Scott Lash, Celia Lury, Mariam Fraser, Sabine Hofmeister, Ida Sabelis, Sierk Ybema and Frederick Vanderberghe. I also want to thank friends outside the academy whose thoughts have greatly influenced me, most notably Emma and Alex van Spijk, Fr Chris O'Connor, Fr Mark Brentnall and Fr Paul Chipchase. Above all, however, I would like to thank my wife and best friend Esther Bolier for her sharp insights and constructive criticism and our children Amy, Mark and Anna, who have endured much during the writing of this book and whose relentless enthusiasm and joy for life has made the sacrifices needed for its completion bearable. This book would not have been possible without their unconditional love and support.

Parts of Chapter 8 have been published as 'Mediating the Risks of Virtual Environments' in Allan, S., Adam, B. and Carter, C. (eds) (2000) *Environmental Risks and the Media*. London: Routledge.

Parts of Chapter 9 have been published as 'Whiter Shades of Pale: Media Hybridities of Rodney King' in Brah, A., Hickman, M. J. and Mac an Ghaill, M. (eds) (1999) *Thinking Identities: Ethnicity, Racism and Culture*. London: Macmillan.

1 Introduction

Technological culture and risk

If anything, the shocking events of 11 September 2001 show that security is a deeply suspect concept. Indeed, we are at risk. It takes only a handful of determined 'terrorists' to cripple the world's only remaining superpower. It takes only a simple failure of 'security' for civic air travel to become part of a war machine and for a passenger plane to become a deadly missile. It takes only a coordinated and targeted effort at a few domains: a financial centre, a centre of government control or a telecommunications system to create chaos and pandemonium. It takes only a few cameras – connected to satellites – to turn such local events into global spectacles with global consequences.

Leaving the spectacle, the apocalyptic ghost of risk returns in mundane and everyday settings. A constant stream of predicted ecological catastrophes with far-reaching implications for our everyday lives have saturated our concepts of the environment since the 1970s (for example, Farvar and Milton, 1973; Taylor, 1970). In health clinics, dozens of leaflets warn us about all kinds of medical hazards we are facing on a regular basis, often without being aware of them. We are told about invisible killers lurking in our ignorance. Blame is allocated to those people who fail to inform themselves about the risks we face. Claims usually follow. Often, however, blaming and claiming are not sensible options. Volatile stock markets wreak havoc in pension schemes, without anyone being held accountable. Indeed, many risks originate from nowhere in particular. They occur without responsible agency. What is worse, the ripple effects of turbulence in financial markets in some distant location may have a near immediate global impact without any possibility of institutional absorption of the shocks.

Taken individually, each risk may have a rational aetiology and can be reasonably explained, anticipated and acted upon. Taken as a cumulative and complex phenomenon, risks appear to be far less reasonable. Taken as a general abstract phenomenon, risks become apocalyptic. It is the accumulation of distinct, often spurious, risks – ecological, biomedical, social, military, political, economic, financial, symbolic and informational – that has an overwhelming presence in our world today. As risk becomes omnipresent, there are only three possible responses: denial, apathy or transformation. The first is largely inscribed in modern culture, the second resembles postmodern nihilism, the third a conversion to some other, not necessarily new, modality of being and thinking.

Sociologists, such as Ulrich Beck (1992, 1995, 1997, 2000) and Anthony Giddens (1990, 1991), have presented ambitious theoretical accounts of the way in which risk has become a force of social change. Both agree that risk challenges the order of modern industrial society and that as risks increase in importance, so does the pertinence of the need for change. Both emphasize the need for social and political reform towards greater reflexivity, taking into account the ongoing processes of individualization which have rendered many traditional and institutionalized response modes to social problems rather dysfunctional if not counterproductive (also see Lupton, 1999).

At the same time, however, and certainly up until 11 September 2001, the idea that we live in a risk society may seem absurd. After all, our world appears a lot safer than that of say war-torn regions or the Middle Ages. Death does not strike us as a random force that operates indiscriminately. We have learned to control a large amount of contingencies, for example related to accidents, violence and disease. Even natural hazards appear less random than they used to be. Although human interventions may not stop earthquakes or volcano eruptions, they can be reasonably predicted. We anticipate them both in terms of structural arrangements as well as emergency planning. The same has often been said of terrorism. Even after the 'act of war against America' as President Bush described it, the prospect of a worldwide war of good versus evil was not perceived to be as ominous and risky as it would have been in the 1970s and 1980s during the Cold War. Indeed, it could be argued that life in modern western societies (but not Afghanistan or Iraq) is now safer than ever.

Though sensible, such expressions however do miss the most obvious point about risk: risks are not 'real', they are 'becoming-real' (Beck, 1992, 2000; Van Loon, 2000a). As soon as risks become real, say an act of terrorism destroying the financial heart of New York, they cease to be risks and become a catastrophe or at least an irritation. Risks have already moved elsewhere: to the anticipation of further attacks, economic decline or worldwide war. That is to say, risks exist in a permanent state of virtuality and are actualized only through anticipation. Hence, it makes no difference whether we are actually or objectively speaking safe; if hazards are anticipated then they call for humans to respond.

In his acclaimed *Against the Gods* Peter Bernstein (1996) argues that what distinguishes modern 'man' from 'his' predecessors is the way in which hazards are being handled.[1] For modern 'man', hazards have to be controlled by 'himself', through systematic application of science and technology, most of all mathematics. That is modern 'man' turns hazards into risks. Modern 'man' does not need God for this. 'He' thus also relates to hazards without the mediation of religion, sacraments or prayer. For Bernstein, risks are human constructs based on both deductive and inductive models.

Bernstein suggests that in contrast, medieval Europeans on the whole did not believe that they could do much about such hazards. They were acts of God (also see Lupton, 1999: 5).[2] However, historical evidence (and this is quite substantial) points towards a much greater allocation of responsibility to humans as acknowledged by medieval writers (Carmichael, 1997; Herlihy, 1997; Horrox, 1994). The biblical Old Testament, of course, contains countless similar stories of how the people of Israel induced God's wrath by disobedience, while being warned of the risks by the prophets. Hence, pre-modern beings were not necessarily as fatalistic as is sometimes suggested. Especially those living within religious systems, based on a concept of a personal God, did have a sense of agency in the face of dangers and calamities. Through prayer and good works, people may appease God and perhaps secure their own salvation as well as that of their families and communities.

What changed in the modern era, however, was that God was gradually taken out of the equation. As a result, there was no longer a supernatural concept of divine power behind hazards. Reason became an exclusive interaction between 'man' and 'the laws of nature' (which were slowly turning into the laws of physics). As a result, 'man's' response to hazards changed. Rather than attuning 'himself' to God's Will (through ritual and doxa), 'man' attuned himself to 'decisions' and systems of 'consciousness'. As Beck (2000), following Luhmann (1995), argues, decisions separate risks from hazards. These decisions usually refer to human actions and interventions.

Hence, there is an obvious direct link between secularization and risk perception. By abandoning God (or even killing Him, as Nietzsche would have it), 'man' has forced 'himself' to provide alternative explanations of calamities, catastrophes and hazards as well as means of regulating them. And this is exactly what modern 'man' has done. If anything, modern technoscience has been strongly motivated by a desire to regulate and secure hazards. By turning anticipation into rational calculation, hazards could be operationalized as risks in terms of probability and this has generated the possibility of 'decision-making' (Krimsky, 1992).

The aim of this book is to explore whether, to what extent and how modern, western society has changed in response to the challenges offered by (increased) risks, especially those induced by technoscience. More specifically, it offers a combined theoretical and substantive analysis of the work of Ulrich Beck by taking it into another domain: that of a Nietzschean and Heideggerian understanding of culture and technology. To do that, we need to call on the help of two distinct but related theoretical approaches: Actor Network Theory and 'biophilosophy'. This is done to show that some implicit elements of Beck's work that have received less attention in recent debates are worth bringing out because they underscore much more clearly the strengths of the risk society thesis. This work is thus not an exegesis of either Beck, Nietzsche or Heidegger, but a selective exploration that focuses on the triangular relationship between risk, technology and culture.

The objectives of this exploration are threefold:

- to provide an introduction to theorizing technological culture and risk by using the work of Beck as a starting point
- to construct a framework for the analysis of relationships between risk, technology and culture based on the risk society thesis, Actor Network Theory and biophilosophy
- to develop more grounded analyses of ways in which the intersections of risks, technologies and cultures operate in the constitution of specific risk perspectives, which for purely practical reasons will be limited to four 'pathologies': waste, infectious disease, 'cybercrime' and (collective) violence.

The central argument of this book will be that the cultural consequences of the risk society have been rather overshadowed by a drive towards a politics of urgency. That is, most responses to perceived risks, including those of denial, are being conceived in haste. The urge to 'act now' is of course itself driven by the acceleration of modernity itself. Saving time by speeding up is the motive behind most technological and organizational innovations (Hofmeister and Spitzner, 1999; Schneider and Geißler, 1999). The politics of urgency, however, inhibits reflection and reconsideration, and as a result ultimately contributes to the proliferation of uncertainties, latent contingencies and thus risks.

However, it is not that risks are being ignored in this process of acceleration, quite the contrary, modern technoscience needs risks to legitimize its claim to scarce resources to speed up its capacity for interventions and to respond more quickly to emergent risks. In doing so risks have been taken up by a form of technological culture in which risk aversion provides the predominant 'ethical' imperative. The emphasis is thus placed heavily on scientific expertise and technological control. The logic of practice of engaging with risks is mainly driven by a managerial approach to regulate these undesired 'bads' and transform them, whenever possible, into 'opportunities'.

The unintended consequences of this narrowing down by means of technological fixation are, on the one hand, excessive surveillance and monitoring that induce a culture of risk aversion riddled with paranoia and neurosis, and on the other hand, an uncanny kind of selective apathy in which major facts are accomplished (as faits accomplis) without any notice. For example, the 'fact' that beef infected with bovine spongiform encephalopathy (BSE) is not 100 per cent safe and may affect humans through consumption was accomplished long before it was finally officially acknowledged. As a result, people who have eaten beef products during the times when the BSE epidemic was at its worst now face much higher risks of possible infection than when they physically consumed the beef. Although even during that time (that is after 1986) there had been quite a few news reports, documentaries and public discussions about possible risks, the UK government, supported by leading scientists, maintained that these risks were negligible or even non-existent (Phillips, 2000; Toolis, 2001). People who believed beef to be safe could have done so on the assumption that experts can be trusted.

Both anxiety and apathy are modalities of risk aversion. Anxiety motivates risk-aversion strategies, apathy is a risk-aversion strategy. Both are extremely prevalent in the risk society today. Ultimately, the risk politics of urgency may be self-defeating. We have to ask whether the duality of anxiety and apathy framing public culture enables us to contain the risks engendered by technological culture. This is unlikely if we consider the growing complexity of risks which exceeds the representational ability of models of managerial planning and control. As more and more complex models of representation and datasets are called for, the predictable effect will be increased volatility, change and disorder. Hence, the risk society is increasingly embracing apocalyptic sensibilities.

The ascendance of risk

During the 1980s and 1990s, the term 'risk' has obtained a pervasive and even intrusive presence in almost all institutionalized discursive fields in modern western societies: technoscience, governance, mass media, economics (including insurance and finance), law and the military. Of course, the term risk was already widely used before in most of those domains. Whether explicitly or implicitly, risk usually appears whenever decisions have to be made.

In societies where 'man' is seen as the measure of things, i.e. those of western modernity, risks cannot but become more prominent. By taking God out of the equation through 'reason', 'man' also eliminates an infinitely unknowable motivation by replacing it with rational explanation, and thus calls upon 'his' own motivations to modify events if they are detrimental to 'himself'. In the late nineteenth century, risk assessments were already essential components of military strategy, insurance business, stock-market exchanges and so on. Risk assessments (albeit crude ones) formed the basis of a range of public health interventions across all western societies in the late nineteenth and early twentieth centuries (Craddock, 2000; Latour, 1988; Lupton, 1994). Indeed, the wide appropriation of the mathematical and statistical concepts of 'probability' (e.g. Adams, 1995; Bernstein, 1996; Prior, 2000) illustrates that the notion of risk has been endemic not only in modern western technoscience, but also in the very institutional infrastructures of twentieth-century capitalist social formations.

Hence, even for modern 'man', risk is nothing new. However, what may be unique is the increasing pervasiveness and intrusiveness of risk assessment in a growing number of social, political and cultural practices. Even more important perhaps is that such generic shifts across discursive fields could signify more fundamental social and cultural changes in how we as humans relate to our own being-in-the-world. This is why risks have become an important area of study for the social sciences and humanities who would otherwise have very little to say about 'probability' and 'risk assessment'.

In sweeping terms, one can formulate the theory of reflexive modernization: at the turn of the twenty-first century the unleashed process of modernization is overrunning and overcoming its own coordinate system. This
coordinate system had fixed understandings about the separation of nature
and society, the understanding of science and technology and the cultural
reality of social class. It featured a stable mapping of the axes between
which the life of its people is suspended – family and occupation. It
assumed a certain distribution and separation of democratically legitimated
politics on the one hand, and the 'subpolitics' of business, science and
technology on the other.

(Beck, 1992: 87)

Since the publication of the translation of *Risikogesellschaft* into English in 1992,
risk has become a significant concept and research topic beyond the limited
fields of technoscience, statistics and mathematics (for example Adam *et al.*,
2000; Allan *et al.*, 2000b; Breen, 1997; Fox, 1998; Lupton, 1999; Macnaghten
and Urry, 1998).[3] There is more than a slight resonance between Beck's ideas
and those of Giddens (1991), whose definition of risk as 'manufactured uncertainties' is central to his understanding of modernization as detraditionalization.
Giddens's work is in many ways central to the development of sociology in
Europe, and this may explain the rapid ascendance of the concept of risk in the
social sciences since the early 1990s. However, other publications may have
contributed here as well, most notably John Adams's *Risk* (1995) and the
American Peter Bernstein's (1996) aforementioned *Against the Gods* whose
perspectives are vastly different from that of Beck and Giddens, as they emerged
from 'rational choice theory' rather than classical sociology. Finally, the work of
Brian Wynne (e.g. 1989) could be mentioned. Whilst instrumental in introducing Beck's work to the Anglo-Saxon world, his own perspective owes more to
social constructionism, similar to those developed in the sociology of scientific
knowledge (Knorr-Cetina, 1981; Knorr-Cetina and Mulkay, 1983; Mackenzie
and Wajcman, 1999; Mulkay, 1979) which had firmly established itself in the
UK as well as continental Europe by the time of the publication of *The Risk
Society*. Wynne's work is particularly interesting because of his emphasis on the
relational character of risk, which is equally central to Beck's thesis.

However, whereas in Europe, the publication of Beck's work is often hailed
as a landmark, there was already an emergent tradition in the United States
dealing with the concepts of risk and hazards as more than a synonym of
probability. According to Krimsky and Golding (1992: xiii) 'the field of risk
studies grew out of the practical needs of industrialized societies to regulate
technology and to protect their citizenry from natural and technological
hazards'. Their collection of essays entitled *Social Theories of Risk* was one
of the first attempts to trace the lineages of more theoretically informed
concepts and paradigms of risks, especially since the 1970s. Susan Cutter's
(1993) *Living with Risk* was similarly hailed as a landmark in opening the study
of risk and hazards for social sciences, in her case human geography, by

providing an overview of perspectives that were implied in particular studies, yet rarely theorized at higher levels of abstraction.

However, the most famous perhaps is Douglas and Wildavsky's (1982) *Risk and Culture* in which risk is being conceptualized in terms of its social function. Douglas and Wildavsky argue that risks operate in a socio-political process, in which the selection and definition of particular hazards as risks has immediate implications for the regulation of social order. The coupling of risk and blame, which Douglas (1992) has followed up in a more recent publication, reflects concerns in her earlier work (such as *Purity and Danger*, 1966) about the regulation of pathologies and the problem of boundaries. By focusing on the social functions of risk, Douglas and Wildavsky were able to shift attention away from the nature of probability calculations, towards the cultural framing of perceptions of risk (Lash, 2000; Rayner, 1992; Scott, 2000). Whereas the centrality of risk perception was already acknowledged by those involved in scientific risk assessments (e.g. toxicology, epidemiology), it was usually understood as related to the interpretation of scientific facts. Douglas and Wildavsky convincingly argue, however, that social, political and cultural processes take precedence over information about scientific facts (Lupton, 1999: 37; Rayner, 1992: 91).

The sociological and anthropological interventions in 'risk studies' have vastly increased over the 1990s. The aim of this book is not to provide an overview of this debate (see for example Lupton, 1999). Instead, it will provide a discussion of aspects of Beck's risk society thesis that have hitherto been relatively neglected, in particular its potential relationships with concepts of culture and technology that have a radically different origin from that of modern sociology. Hence, it will discuss the concept of risk in relation to two strands of contemporary social and cultural theoretical work: Actor Network Theory and biophilosophy. In the background of these discussions are two philosophers whose works will hardly be discussed as such, but who are silently present even if in rather distorted forms (for this is not a book on philosophy): Nietzsche and Heidegger. None of these figures has written anything about risks as such. However, if Beck is right that risks do mark a significant shift in the social, political, economic and cultural organization of society, then we must be able to trace parallel claims in these seminal and influential theories. This book hopes to show that reflections by other theorists on how to understand such changes are of paramount importance to understanding where risks are taking us. The purpose of this is to develop new ways of conceptualizing risk by synthesizing a selected range of theories and related concepts.

Risk, technology and culture

This book aims to conceptualize risk in relation to two other concepts: technology and culture. The terms technology and culture are often used too loosely and casually in political, popular as well as academic discourses. However, a proper exploration of these concepts is part of the whole process of analysis, not the introduction. Hence as a shortcut, that is, before the more detailed discussions of subsequent chapters (and especially Chapter 5) the concept of technology used here is partly derived from Heidegger (1977) as the way in which our world is 'ordered' and 'revealed'. Ordering refers to both 'putting into place' (e.g. classification or enframing) and 'commanding' (giving orders). Revealing means 'bringing into existence' (making appear). In a similar simplification, the term culture is used as a mundane and everyday set of practices of sense-making which entail both 'inhabitation' or cultivation (as in agriculture) and 'signification' as the attribution of meaning (Van Loon, 1996a). Again this owes much to Heidegger's (1954/1991) philosophy, although Nietzsche's influence, via poststructuralism and semiotics, is equally important.

However, there has been a convincing and sustained criticism of Heidegger's rather deterministic concept of theory, which cannot be ignored. Feenberg (1991), for example, criticizes substantive theories of technology such as Heidegger's and Jacques Ellul's for failing to consider the social, cultural and political embedding of technological change. Rather than being an autonomous force, external from and imposed upon 'society', Feenberg (1991: 13 and Chapter 8) argues that technology does not necessarily engender alienation, but can be effectively challenged from within, as it is also part of a political (class) struggle. Inspired by Lukács and Marcuse, he thus offers a critical theory of technology to advocate a radical approach to technology as part of a general strategy of progressive democratization.

Although in the face of the risk society, such a view may turn out to be far too optimistic, it does serve as an effective warning against a one-dimensional approach to technological culture. As contributors to Bijker and Law (1992) and Mackenzie and Wajcman (1999) have also showed, the relationships between technology and culture are far more complex than can be accounted for by technological determinism. What matters most here are the relationships between the concepts of risk, technology and culture. It is the juxtapositioning of these three that generates the perspective informing this book. Technological culture refers to modes of inhabitation and signification that are not simply using technologies but are also produced by technologies. The term enables us to theorize more explicitly three rather well-established but unhelpful problematics that have structured many sociological debates about technology and culture:

- technological versus cultural (or social) determinism
- the role of technology in modernity
- the status of the human being.

As already mentioned, suggestions that technological change enforces social change are often accused of suffering from technological determinism. It is often pointed out that existing human relationships, established meaning formations, everyday practices, interactions and social structures shape technological changes (Kline and Pinch, 1999). That is, technological innovations are often a response to perceived social, economic, political or cultural needs. However, it is also generally accepted that the mirror image of technological determinism, e.g. social or cultural determinism (or instrumentalism) is equally flawed. Technologies cannot be reduced to mere responses to human needs for this fails to adequately comprehend the additional and unforeseen consequences of many technological innovations beyond the intended purposes for which they were designed (Feenberg, 1991).

Hence, we need to problematize both instrumentalism or cultural determinism and technological determinism, but without making the opposition disappear in a sublation. The tensions between technology as an instrument (tools) and as a force are not fictions of speculative imagination. They are activated in our ordinary everyday lives: in decisions over how and what to eat, what to buy, how to go to work, how much television children are allowed to watch and so on. On many occasions, we approach technology through tools. Appropriating technology in this sense seems completely inconsequential to our being. We use either a microwave or a gas cooker, a drill or a hammer, a personal computer (PC) or pen and paper. However, these appropriations are far from inconsequential. They entail different mobilizations of time, energy and costs, and – equally importantly – they reveal different realities to us, for example, about the nature of food, walls or writing.

A second consequence of the juxtapositioning of technology and culture operates at a more general level. The term 'technological culture' highlights that the modes of inhabitation and signification (culture) that make up our world are technologically mediated. Whereas to some degree this is the case for most civilizations throughout history, there is something distinctively *modern* and *western* about technological culture. The 'modern' element comes from a fundamental rupture in the relationship between history and human being (Foucault, 1970). This rupture started in the mid-eighteenth century, with the advent of industrial capitalism, the nation-state (especially after the French Revolution) and technoscience (Stiegler, 1994: 51–3). It first reversed the relationship between 'man' (again used with full patriarchal connotation) and nature. Technological innovation enabled a domination over nature not seen before. Especially the steam engine granted greater speed, mobility and power to 'man' to overcome natural obstacles more efficiently. This was quickly converted into more elaborate technologies of extraction. Nature was increasingly being turned into a resource or, as Heidegger (1977) has argued, a 'standing-reserve' (see Chapter 5). Although technological innovation has a much longer lineage, it was the coupling with science that enabled the reversing of the relationship between nature and 'man' in favour of the latter (also see Stiegler, 1994: 58).

Science itself underwent a significant break. As Foucault (1970) has argued, this break entailed an epistemic shift from thinking in terms of 'classification' to that of 'causation'. Causal logic enabled science not only to predict what would happen following particular interventions (after all, this was already done by experimental empirical science of the sixteenth century), but also to explain them logically *beforehand* on the basis of principle rather than empirical observation. Experiment and observation were thus turned into arts of verification rather than discovery which, in turn, enabled 'man' to explain the very logic of life and ultimately 'his' own being; this was the birth of human sciences. When Nietzsche wrote that 'God is dead and we killed him', he was referring to the displacement of the mysterious and sacramental nature of the Divine by what for him always-already came before it: the clever beast called 'man' (Nietzsche, 1983). Hence 'man's' domination over nature was extended with 'his' domination over God (Nietzsche, 1990).

Whereas a full-scale secularization of western society would not take place until the second part of the twentieth century (almost a century after Nietzsche's bold proclamation), practically speaking, science and technology had already made religion quite redundant. For modern thought, institutionalized religion was simply slow in recognizing that for modern western society it is a matter of logical extrapolation that 'man' and not God determines the agenda and gives the orders.

With 'his' dominion over nature and God firmly established, 'man' could now also claim a new relationship with history itself. 'Man' was now both origin and destiny of 'his' own fate. 'Man' was both the measure and the ultimate benefactor of all things. This was called progress. Progress entailed the historical becoming of 'man', through 'his' emancipation from both Nature and God. In modern culture 'man' is destined for progress; progress is the destiny of 'man' (Foucault, 1970: 344). Technological culture must thus be understood as a manifestation of modernity. 'Man's' relationship with 'his' being in the world is mediated by technologies, that enable 'him' to subordinate that world to 'his' own becoming.

This immediately signals that technological culture is also very western. Not only because modern technoscience to a large extent originated as a western phenomenon, but also because its political and economic infrastructure is deeply entrenched in western capitalism and western nation-states and their international institutional structures. However, rather than the highly problematic term 'western culture', technological culture does not assume a geographically bounded homogeneity of either values or ways of life (or both). Although the nodal points of technological culture are usually traceable to 'the West' (including Australia and Japan), many urban centres around the world are marked by the considerable globalizing influence of technological culture (Simone, 2000). And whereas technological culture is certainly framed on the basis of 'western' interests, and in many ways is geared to serve these, its consequences certainly are not restricted to 'the West'. Moreover, the western pejorative of technological culture cannot prevent its (sometimes) deviant appropriations by others against its original script (Friedman, 1990).

Alongside the breaking with the opposition between cultural and technological determinism, and highlighting the historical and cultural contingency of technologically mediated 'human' being, a third important feature of the juxtaposition of technology and culture is its problematization of human *agency* (or will or consciousness) as universal and unique. When the possibility is acknowledged that cultures can be technological, then we must also be open to questioning what kind of cultural agents may actualize it, other than humans alone.

Agency is generally understood as synonymous with the ability to act according to one's will. Its significance for modern thought cannot be overestimated as it is in the notion of 'agency' that humans could see themselves as the makers of history. The main question for modern social sciences has always been to what extent agency is not already informed by structures (Giddens, 1984), both in terms of constraints (especially in the form of law, restrictions, regulations) and access to resources (economic, symbolic, political, social, spatial). However, even for the most restricted concepts of agency such as those found in structuralism, the main tenet has always been that agency is in principle connected to free will. That is, agency is the ability to act on the basis of consciously made decisions.

However, this human-centred notion of conscious agency is beginning to lose its persuasive power. Even if we do not accept the argument that machines have a consciousness, it is possible to argue that they can act without human intervention, as if they were completely autonomous. The important issue here is whether consciousness (or free will) is a necessary precondition for agency. At this stage of the argument, we have to be agnostic and allow both arguments to coexist. That is, agency requires free will, but the ability to act is also independent of consciousness. It is not simply a matter of opinion (assertion) but of logic and argumentation which can be traced back to the work of Nietzsche (Ansell Pearson, 1997) and even further perhaps, to Parmenides in particular his concept of *pathos* (e.g. Despret, 1998). However, even if we wish to distinguish between agency (based on free will) and the ability to act (in relation to which the term 'actant' will be introduced in Chapter 3), the concept of technological culture enables us to understand culture as more than human, and the generation and organization of meaning as more than a human pejorative.

Information and communication technologies

By addressing risks in relation to technological culture, the connections between the political-economics and socio-cultural formations of 'modern western societies', including its global and historical context and the actualization of (moral, political) agency, become the main foci of attention. Alongside the steady advance of media technologies such as print and broadcasting, the rise of information and communication technologies (ICTs) must receive particular consideration here. Irrespective of whether one understands risks in a realist or constructionist framework (see Lupton, 1999), mass media and ICTs play a major role in the formation of risks, risk sensibilities and risk perceptions (Van Loon, 2000a). It is remarkable, however, that until recently, very few studies have concerned themselves systematically with the role of media in the formation of risk perceptions. Although the media are often included in discussions of the public understanding of science, and are seen as central to the dissemination of public information about hazards (e.g. Cutter, 1993; Hannigan, 1995; Ungar, 1998), very little work actually engages with a less instrumental and more productive perspective of media as forces of risk themselves. Exceptions to this include the work of Cottle (1997) and more recently a collection of essays edited by Allan *et al.* (2000b). They suggest that any concept of a risk society relies heavily on a theory of mass mediation (also see Lupton, 1999). This means that rather than engaging in a critique of media representations of risk, we should be focusing on the constitutive role of media in the production of risk. That is, media are part of the technological constellation through which risks come into existence.

As the global economy, the world political order and most socio-cultural systems are nowadays bound by high-speed and high-frequency information flows, there is no escape from the impact of telecommunications on processes of decision-making and anticipation. However, apart from accelerating information flows, ICTs also contribute to the acceleration of risks. One of the main consequences of the compression of time to immediacy is that contemplation and reflection get more and more squeezed out of decision-making processes. This has a profound impact on the nature of social and political organization. The growing immediacy of information and communication reveals a world that is increasingly a fait accompli, a world always-already 'there'. Rather than through decisions, technological processes order a world through what Lyotard (1979) referred to as 'performativity'.[4]

Performativity (enactment) is the instrumental logic of optimizing the processing of inputs into outputs. It is entrenched in the nature of technology as 'ordering'. This 'optimization' follows the simple logic of reducing costs and maximizing benefits. It is thus inherently linked to a drive towards greater efficiency and effectivity because it seeks to minimize the difference between thought and action to the point that the two are completely united. When the instantaneous speed of thought becomes the immediacy of accomplished action, technoscience is no longer dependent on discourse and representation in effectuating man's full mastery over nature. When that has been accomplished, there is no need for politics or reasoned deliberation. The medium becomes obsolete. Indeed, what use is there for political and moral intervention by human and perhaps non-human actants to engage deeper senses of truth and justice than those already set to work by performativity?

The telecommunications revolution provides an excellent illustration of both McLuhan's (1964) visionary account of the role of media technologies in the transformations of the human condition, and Alain Touraine's (1971) equally visionary but far more pessimistic observation that our world is becoming increasingly 'technocratic'. Though the prophecies of McLuhan and Touraine may seem outmoded, their continued relevance becomes obvious when considering the works of contemporary writers such as Manuel Castells, Mark Poster and Paul Virilio on the impact of digital information and communication technologies on social order and social change.

Against the frequently endorsed complaint that such analyses present forms of technological determinism, it can be argued that each of them understands technologies to be embedded in historically and culturally specific, social, economic and political relationships. What they do object to, however, is (a) the suggestion that technological change is an intentionally produced outcome of social struggles and (b) the idea that technology is merely instrumental in the deployment of agency (the will to power as 'free will'). Of course, struggle, warfare, and revolutions matter a great deal, as indeed many technological innovations have emerged from military experiments (Virilio, 1986); however, it is a mistake to conceive technologies, or social relationships for that matter, as exclusively human products that are solely governed by human needs and intentions. Human beings are themselves produced by a particular technological culture: that of 'modern thought' (Foucault, 1970). Therefore, rather than embracing either misanthropic technological determinism or what may be called an 'opportunistic humanism' (a coupling of instrumentalism with a naive belief in linear progress, even if this includes a 'post-human' successor), we need to engage more critically with 'technological culture' as a phenomenon that is intertwined with the ordinariness of life, i.e. as 'inhabited'.

Rationale

The book consists of two main parts. The first half of the book deals with theory. The aim is to get a general idea of how to theorize the technocultural framing of 'risks' in contemporary society and its possible consequences. The approach followed here is an attempt to develop the risk society thesis in a new direction in which technology and culture are primarily seen as everyday practices rather than abstractions.[5] This leads us to make specific connections between Beck's work and theories that have been more overtly influenced by such thinking. Hence, in addition to Risk Society Theory (Chapter 2), Part I also incorporates Actor Network Theory (Chapter 3) and biophilosophy (Chapter 4). Chapter 5 provides a conclusion to Part I and attempts to reconstruct a theoretical framework with which to approach the risk phenomena that feature in Part II. It will show that what links the risk society thesis with Actor Network Theory and biophilosophy is an implicit relationship to a philosophical ethos inherited from the works of – but not exclusively – Heidegger and Nietzsche. The Heideggerian 'turn' will be explored in more detail in Chapter 5, and enables us to understand the significance of the concept of risk beyond sociological abstractions.

Part II presents four illustrations of particular forms of risk related to ecology, in the form of waste (Chapter 6), public health, with a specific focus on emergent infectious diseases (Chapter 7), information and communication technologies or 'cyberrisks' (Chapter 8) and collective violence, specifically urban 'riots' (Chapter 9). The conclusion will draw these chapters together in order to refine the analytical framework and suggest ways in which we could proceed, in terms of research, theorizing as well as political action.

The first objective of this book is thus to provide an introduction to theorizing technological culture and risk by using the work of Beck as a starting point. Beck's work is often understood as being part of a particular tradition within contemporary sociology. This tradition maintains a rather classical conceptual framework for understanding the nature of social order and social change; one that is based on a triad of (in Beck's case) pre-modernity, industrial modernity and 'another modernity'. Risk society is the constellation of forces which propel industrial society into disarray, cultivating the necessity of the reinvention of a (global) politics against the instrumental rationality of industrial modernity. However, such a focus on industrial versus risk society obscures Beck's rather more fascinating insights into the way in which risks are culturally engendered (and involve media technologies). There is a possible cultural-theoretical dimension to his work that hitherto has been neglected by his main critics. One of the objectives of this book is therefore to explore the cultural aspects of Beck's risk society thesis. What sort of sense-making practices are involved in the cultivation of risk perceptions and sensibilities? In order to answer this question, it will be necessary to extend from Beck's work into theories that have been developed by a few of his contemporaries.

Actor Network Theory, and especially the work of one of its proponents, Bruno Latour, will be discussed as a very useful way of conceptualizing the practice-based technoscientific formation of risk. Although Latour himself rarely mentions risk, his highly perceptive understanding of how science constitutes its object can be easily applied to this elusive concept. Actor Network Theory can be developed to argue that the constitution of risk as an object, i.e. its objectification, involves a complex process of making appear (visualization), attributing meaning (signification) and allocating value (valorization). Central to Actor Network Theory is that humans are not the only actors. This is also an essential axiom of technological culture.

This will be further theorized in Chapter 4 with reference to what has been termed 'biophilosophy'. Biophilosophy is an amalgam of a range of theoretical perspectives loosely associated with poststructuralism, postmodernism and complexity theory. Although by no means coherent enough to deserve to be called a paradigm or perspective, biophilosophy is recognizably marked by a shared desire to redefine 'life', to break with the anthropocentric concept of organic integrity (which still prevails, even in current forms of Darwinism), and radicalize a concept of life processes as 'flows'. In doing so, it generates a rather uncanny perspective on risk, in which it is no longer seen as a threat to integrity, but as a necessary point of departure, a beginning of new life. Although the list of biophilosophers is large, the chapter mainly focuses on a few 'usual suspects': Donna Haraway, Gilles Deleuze and Felix Guattari and Jean-François Lyotard. This is not to deny that others are less relevant. However, as this is not a book on biophilosophy, we cannot but be selective.[6]

There is of course already a strong selective element in juxtaposing risk society theory, Actor Network Theory and biophilosophy. Whereas most selectiveness is rather arbitrary, some of it does have a logical basis. Their successive ordering is used to tell a certain story. This story is the second objective of this book. It attempts to understand risk in relation to the current human predicament. This story has been told in parts by many different philosophers and theorists, among them Nietzsche, Heidegger, McLuhan, Ellul, Touraine and Marcuse. Whereas this is also the central focus of Risk Society Theory (hence our point of departure), it will be made clear that macro-sociology is rather inadequate in understanding the nuts and bolts of the mechanism of its own argumentation. By integrating elements of the thesis with theoretical work that starts with different sets of assumptions, we will see that the common thread that unites some of the insights thus generated has a logical basis that is inherent to modernity itself.

However, the book intends to go beyond providing mere introductions to particular theories. The third objective is to use the analytical framework to make more sense of the way in which risks are being revealed, ordered and inhabited by technological culture. The intention here is to go beyond the critique of modern technoscience as both a production and a negation of risks. This critique is problematic because it prevents us from understanding the fundamental role of risks in the development of particular technoscientific projects. Scientists are all too aware of the intimate relationships between technologies and risks; it is on the basis of this intimacy that they can develop actor networks (e.g. Wynne, 1996). The very idea of (enhancing the) 'public understanding of science' stems from an anxiety over the fragility of these actor networks (Irwin, 2000; Rose, 2000).

Instead of a moral critique of science and technology (as a 'mode of rationality' or 'episteme'), this book aims to provide an immanent but critical engagement with science and technology as *cultural practices*. Although theoretical rather than ethnographic, it will be argued that understanding risks requires a thorough understanding not only of the technologies of risk production, but also the technologies of risk management in the context of their ordinary everydayness, rather than spectacular manifestations in catastrophes.

Part I

Theoretical framework

2   Cultivating risks

Paradoxes in the work of Ulrich Beck

This chapter presents an analysis of Beck's risk society thesis and in particular its often neglected cultural dimension. It will look at the main thrust of the analysis, which is focused on the transformative consequences of 'risk' in late-modern societies. Particular attention will be given to the complementary but paradoxical processes of individualization and reflexive modernization. The centrality of the figure of the paradox is evident not only in Beck's style of writing, but also in his account of the risk society itself. The main paradox which will be drawn out here is that between the increasingly central role of technoscience in identifying and managing risks and the growing delegitimation of science and technology as a result of the failures of containing these risks. The failure to focus on Beck's paradoxicality has led many commentators to accuse him of being both too optimistic and too pessimistic. The aim of this chapter, however, is not to provide an exegesis of Beck's oeuvre, or even his writings on risk, but to use some of his ideas to elicit a more in-depth understanding of the significance of 'the risk society' as arguably the most adequate description of our current technological-cultural predicament.

The paradox also pervades Beck's sense of the political. His critique of the 'subpolitics' of technoscientific expertise (Beck, 1997) can be associated with Habermas's theory of Communicative Action. Indeed, Beck shares with Habermas a distrust of instrumental rationality as well as a commitment to thinking ethics in terms of 'justice'. However, to understand Beck's paradoxical writings more fully, we must also emphasize the implicit legacy of Niklas Luhmann in his work. This allows us to see why Beck is reluctant to simply embrace either a dystopian-revolutionary or utopian-emancipatory ethos, and is more inclined to explore and deconstruct the self-referential character of such concepts and the call-to-arms these sensibilities generally implicate.

Excess modernization

In the *Risk Society*, Beck (1992) puts forward the argument that the developments of science and technology have generated major problems for industrial society, which he treats as synonymous with 'modernity'. As is common in the sociological tradition, Beck perceives the start of modernization in terms of the birth of industrial society. He combines ideas from Marx (1984) and Weber (1956) to describe industrial modernity as capitalist and rationalist (Sayer, 1991). However, less visible but equally essential in his work is the uncanny implication of Durkheim's main contributions to understanding cultural modernization in terms of differentiation, anomie and the *conscience collectif*. It is because of this that some commentators have labelled Beck's work as (critical) 'structuralist', providing a critique 'of the ways in which social institutions . . . wield power over individuals, reducing their capacity for agency and autonomy' (Lupton, 1999: 26). This label, however, is quite misleading because it ignores the key function performed by paradoxes that pervade throughout Beck's work. As will become clear, the paradox disrupts the neatly organized one-dimensional categories of structuralism versus subjectivism and constructionism versus realism.

It is important to first get a grip on what Beck refers to as the first stage of modernity or industrial modernity. In industrial modernity, the primary social relationships and antagonisms are defined in terms of class. Although Beck does not stress explicitly that class relations are an inherent part of the capitalist mode of production, traces of Marxist theory are present in his assertion that classes are based on a differential ability to satisfy needs (Beck, 1992: 44). Ability must be understood here as more than having access to resources (wealth), as it extends into controlling the conditions of production and reproduction. Industrial society is primarily based on the production and distribution of goods, for which scarcity functions as the main regulatory principle. Needs, defined in terms of scarcity, are usually actualized in a direct association with 'goods'. They are therefore visible and often seem very real.

The problem with a social system based on scarcity and visible needs is that the institutions that have been created to regulate the production, distribution and redistribution of goods can only deal with their manifest actualizations, not with their latent or virtual potential and side effects. In the era of industrial modernization, the visible has dominated the invisible (Adam, 1998). The institutions of industrial society have not been designed to handle and process the production and distribution of the invisible goods or 'bads' – which Beck predominantly conceptualizes as the risks and hazards that emerge with industrial production. And it is the paradox of risk that it thrives in absence.

> In the overlap and competition between the problems of class, industrial and market society on one side and those of the risk society on the other, the logic of wealth production always wins, in accordance with the power of relationships and standards of relevance – *and for that very reason the risk society is ultimately victorious* . . . At a certain stage of social production . . . the predominance of the logic and conflicts of wealth production, and thus the social invisibility of the risk society, is no proof of its unreality; on the contrary, it is a motor for the origin of the risk society and thus a proof that it is becoming real.
>
> (Beck, 1992: 45)[1]

The main argument Beck thus puts forward is that whereas these risks and hazards have been the undesirable latent side-effects of industrialization from the beginning,[2] it is only with their manifestation, for example in the early 1970s 'Limits to Growth' debate, that they have started to undermine the institutions of industrial modernity. As long as risks are secondary to scarcity/needs, the institutions of industrial society will contribute to their proliferation. However, somewhere in our relatively recent past, a qualitative change occurred in the perception of social order as based on flows of goods *and* 'bads', rather than goods alone. This change in perception has led to a crisis in the way in which modern western society organizes and manages its institutions and functions. This crisis, it could be argued, is the transition period between industrial modernity and something else. The interregnum is referred to as the (world) risk society (Beck, 1996).

The advantage of placing risk at the centre of changes in social organization is that risks can no longer be treated as mere side-effects of industrial society. In this sense, Beck differs from many of his contemporaries such as Adams and Giddens. If risks are predominantly seen as side-effects, this merely perpetuates the predominant institutional response of a pre-reflexive risk society, further contributing to the handing over of the management of 'bads' to the sovereignty of expert systems. In other words, and in contrast to what has been claimed by some of his critics (Alexander, 1996; Levitas, 2000; Scott, 2000), Beck is arguing neither that industrial society was less risky, nor that the risk society has simply replaced the industrial society. Scarcity still matters and social class is still relevant, especially when considering exposure to a wide variety of (often industrial) risks (Bogard, 1989; Cutter, 1993; Eyles and Woods, 1983; Hazarika, 1987). However, what is different is the relative relevance of scarcity in comparison to risk, and class relations in comparison to risk relations.[3] That is to say, the first paradox, that risks proliferate when they are not articulated, is supplemented with a second paradox, that when risks are articulated they change the conditions of their own reproduction. They do so by inducing an environment that is very congenial to their replication: an environment in which risks transform anxieties into blame and hazards into market opportunities. This induction process primarily evolves through the production of knowledge and the dissemination of information. 'The risk society is in this sense also the science, media and information society' (Beck, 1992: 46).

The production and dissemination of flows of 'bads' has always been part of the systems of production and dissemination of goods. Indeed, 'bads' function as necessary supplements of goods. In 'industrial modernity' the primary 'bad' is '*not-having*' or scarcity. For Marx, not-owning means of production forced workers to sell their labour power, which is the first step in the process of alienation (Marx and Engels, 1988). In a similar vein, Weber's (1985: 221) notion of the *Entzauberung der Welt* (the elimination of magic from the world) was based on the assertion that the mystical was merely a lack of reason (fideism), an obstacle to be overcome by rationalization which was the inevitable drive of the modern state apparatus (Giddens, 1985; Lash, 1990). In other words, the motivation behind modernization has always been the same: the overcoming of lack. Progress and emancipation are the driving forces of modernization, scarcity is its main obstacle (Sayer, 1991).

Yet, were it not for scarcity, one could not conceive of modernization at all. Capitalism thrives on lack (translated into need), which urges people to work and consume; law emerges out of lack in regulation; science emerges from a lack of knowledge, etc. The coupling of lack and modernity is axiomatic in theoretical systems as diverse as liberal economics and psychoanalysis. In these systems of thought, scarcity and lack are 'bads' that generate goods through mobilizing desires. Two conflations take place at once: first, the positivity of goods is marked by the negativity of bads; second, this positivity itself equates productivity and moral goodness. Indeed, such 'productivism' is most fully embodied in the Calvinist ethic in which the calling, or vocation, finds its ultimate materialization in productivity and fruitfulness. 'Not leisure and enjoyment, but only activity serves to increase the glory of God, according to the definite manifestations of His will' (Weber, 1985: 157).

For example, when in the nineteenth-century municipal governments in Paris, London or San Francisco articulated the discovery of a possible relationship between poverty and tuberculosis, the pathogenesis of the disease was exclusively phrased in terms of a lack of adequate provisions in terms of sanitation, adequate diet, ventilation and space (Craddock, 2000; Latour, 1988). What such accounts did not stress explicitly, but none the less assumed, was the presence of a bad that is not a lack. It was only with the ascendance of germ theories such as those of Koch and Pasteur and associated technologies of vaccination and pasteurization, however, that this lack could be made present as a bacillus (Wildey, 1987).

One may of course argue that it does not make much of a difference whether the bad was a bacillus or a lack of adequate housing and health. However, when this is compared with the outrage caused by the inaugural speech of South Africa's President Mbeki at the Thirteenth World AIDS Conference in Durban on 9 July 2000 when, extensively quoting from a World Health Organization (WHO) Health Report of 1995, he suggested that poverty should be seen as a prime target in the fight against AIDS (acquired immune deficiency syndrome), the difference becomes clear. After summing up a long list of diseases threatening the African continent, President Mbeki asked 'are safe sex, condoms and anti-retroviral drugs a sufficient response to the health catastrophe we face?'.[4] The angry response directed at Mbeki was that he failed to single out the human immunodeficiency virus (HIV) as the *single* killer and dared to say that the biggest threat to the African continent was extreme poverty, under which he *subsumed* AIDS. He was rebuked for daring to question the wisdom of allocating *most* resources to the promotion of safe sex, even if he pledged support for this. In the age of AIDS, to speak against the HIV-hypothesis is to make oneself a target for ridicule and strong, often emotive, disapproval. For the global HIV/AIDS lobby, it is an article of faith that the virus is the bad, not the lack of goods (except, perhaps, condoms).[5] The difference between goods and bads matters.

The change in appreciation of how bads are to be articulated in relation to a politics of urgency is the crucial turning point in which the risk society is inaugurated. The politics of urgency remain, bads still need to be overcome, but they are no longer primarily defined in terms of absence of goods. In a perspective of industrialism, bads are simply goods-in-minus that is, they are simply the inversed mirror image of goods. However, the perspective of the risk society disrupts this binary order. Goods and bads no longer operate on the same scale. Alongside scarcity a whole new ensemble of 'bads' have emerged. Of course, late nineteenth-century medics were fully aware of the works of Koch and Pasteur and inundated in germ theory (Smith-Hughes, 1977; Van Loon, 2002). They knew very well that certain pathogens can cause infectious diseases.[6] However, these pathogens were seen as part and parcel of a natural world that was still to be conquered by pasteurization, vaccination and later antibiotics. There was no concept of these pathogens being themselves part of wider productive assemblages of modernity. Their total eradication was the aim of medical science.

In the risk society, this is no longer an option. This is because bads are not exclusively seen as obstacles still to be overcome, but as manufactured side-effects of this very attempt at overcoming. The worldwide dissemination of HIV became possible only in a modernity in which the speed of movement has exceeded the mobility resistance of infected/infectious bodies (Ewald, 1994; Wills, 1996).[7] Indeed, within the average incubation time of Ebola, one of the deadliest diseases known to man, one can travel around the world several times, have physical contact with thousands of people and disappear without a trace, that is, until the pathogens reveal themselves. With HIV or new variant Creutzfeldt–Jakob disease (vCJD), whose incubation times are years rather than days, the opportunities of spreading are vastly enhanced. Apart from travel and transportation, infectious diseases are also engendered by other modern technologies such as electricity dams, the destruction of rainforests, agricultural innovations, pesticides, the industrialization of food production, genetic engineering and even western medicine itself (Farvar and Milton, 1973; Garrett, 1994; Shope and Evans, 1993; Wills, 1996). The number of new pathogens has rapidly increased since the 1970s (Morse, 1993; Ryan, 1996; World Health Organization, 1996). This is perhaps due to better and more precise diagnostic technologies, but perhaps also because of modernization itself (see Chapter 7).

We now live in a world in which there is an excess of bads. These bads are in excess of the goods that were produced to counteract other bads. The bads are themselves produced by a modernity that has become excessive. In the risk society, bads no longer complement goods, they *supplement* them. That is to say they add to but do not add up.[8] They do not produce a homeostatic or even stable balance; instead they displace this balance. We can only speculate whether this displacement is ever increasing, as in medieval apocalyptic visions, or temporary. The latter seems to be more in tune with Beck's risk society thesis. The supplementary relationship merely indicates that alongside the goods have come bads that are not contained by the existing systems of production and reproduction. The institutions of modern society, which are still mainly framed on the basis of industrialism, scarcity and the moral goodness of productivity, are inadequate to deal with these supplements because they can only produce more of them, and it is not difficult to see that the longer it takes to overhaul this logic, the more realistic the medieval apocalyptic visions will become.

Beck's risk society thesis could thus be read as a warning, as an incentive to mobilize a political movement that works against the productivism of industrial society. In the age of HIV, there is indeed a sense of urgency attached to developing strategies beyond the reduction of scarcity. A more equal distribution of resources on its own will certainly not stop the virus from spreading. Rich people die of AIDS as well. However, it is equally questionable whether the technological fixes – be it those developed by the pharmaceutical industry or by contraceptive education – will fare any better. For example, a series of in-depth ethnographic studies of the early years of AIDS in Marseilles by Bernard Paillard (1998) indicate that the proposed technical fixes of condom use, syringe supply or exchanges, or out-reach work targeted at so-called high-risk groups, have had only limited impact on preventing the spreading of the disease (also see Bloor, 1995). Despite millions of francs having been put into the development of inter-institutional coordination, awareness campaigns and of course scientific research itself, the rate of HIV infection still increased throughout the city.

The technological fixes, themselves products of industrial modernity, cannot stabilize the flows of bads on their own because they merely recycle, fragment and disperse them across a multiplicity of technoscientific factions. This is because, on the one hand, the technological fix assumes a strategy of social engineering based on the full transparency of cognition and reason, thereby reducing the entire process of behavioural change to a simple matter of communication and awareness. This, of course, is increasingly problematic in a world relying on continued specialization and compartmentalization. On the other hand, the technological fix targets a specific bad (e.g. HIV) with a specific good (e.g. condoms or azidothymidine (AZT)) within a regime of 'safe sex'. It thereby ignores the excessive nature of the bad itself. It targets one domain, leaving the rest open for the further amplification of risks. A retrovirus such as HIV is not merely an assemblage of proteins, it deploys intelligence, develops strategies, it responds, acts, mutates and simulates. Indeed, it is a particle that opens up the equations and disables probability assessments. It magnifies risks simply because this type of bad does not seem to have a complementary ('complimentary') good.

Perhaps a too forceful differentiation between the bads of industrial and risk society leads to an overly pessimistic account of the future. Perhaps the distinctions are more a matter of perception than reality. Indeed, it must be stressed that Beck (1996) often endorses such a distinction to highlight how perceptions have changed. However, he also argues that it does not stop with perceptions. His account cannot be reduced to a history of ideas. The experience of HIV *is* different from that of tuberculosis. This may be because beyond this experience lies a material and physical reality that limits the possibilities of becoming-real to a set few. Microbiologists and virologists would certainly agree: bacteria (tuberculosis) and viruses (HIV) are completely different types of pathogens. However, we do not need to go as far as to speculate on the internal relations between objects and perceptions. For a sociological account of risks, it suffices to refer to experiences and their technological mediations (Hacking, 1983; Knorr-Cetina, 1983; Latour, 1987). The central argument in Beck's thesis is that in the risk society, the nature of experiences of bads has changed dramatically and fundamentally, and this has profound and far-reaching consequences for social organization.

Individualization and reflexivity

The first paradox of the risk society is that it thrives in obscurity, that is to say, the more its existence is being denied, the more it *takes place*. The second related paradox is that risks inherently change the conditions of their own reproduction, that is to say, part of the logic of their production is a displacement of what made them appear in the first place. This means that their ontology is inherently unstable; it cannot be fixed. In the risk society, risks manifest themselves as particular symptoms of a virtual cause. They attract particular responses that – often latently – also modify the virtual causality of the risks themselves. This is not to say that all risks do this, but that a certain, even increasing, number of risk types do. For example, nuclear risks, genetic risks and risks of infectious diseases all become visible in symptomatic effects whose causes are only speculatively representable via technological mediations that are not independent from their initial (theoretically informed) revelations as 'atom', 'gene' or 'virus'.

Another word for this problem is complexity. Due to the complicity of technological mediations, the symptomatic phenomena of risks interact with their virtual origins to produce a second order virtual-symptomatic system. Complexity does not mean that specific risks cannot be contained, but simply that this containment may always also engender new risk disclosures of a more complex nature. This is what is referred to in the previous section as excess modernity – there always remains something 'bad' after the attempts to stabilize the flows of goods and bads. This supplement, or displacement, is not beyond our experience. It manifests itself, for example, in suspicions and doubts about technological fixes, scientific certainty, governmental regulation, accurate mediations and profitable business opportunities. All these 'expert systems' of modern society have been forced to surrender more of their previously unchallenged claims to authority. This is not to say that their authority was never challenged before, but simply that such challenges have become normal rather than exceptional. The effect of these challenges is an increased disembedding of individuals from positions assigned to them by these authoritative institutions within the structures of modern life.

For Beck, the positions which anchored subjective experiences into the systems that structure and organize modern life have become destabilized. Class is a major example of this.[9] It is derived solely from the relations-of-production which manifest themselves primarily in the workplace. However, as Touraine (1971), Gorz (1982) and others have shown, work can no longer be understood as organizing the logic of modern experiences on its own. It has been supplemented by other spheres of life. This process has not been smooth. The history of labour movements shows the unease caused by a loss of hegemonic potential of class politics (e.g. Wieviorka, 1995). Even in the UK, a country in which class still is one of the most defining social entities of the political and economic landscape (certainly compared with other western countries), class articulations are now primarily functional in cultural terms, and in many cases disrupted by other inequalities such as those of gender and ethnicity. Only the staunchest and most uncompromising structuralist Marxists can now maintain that class politics are in essence still universal. That is, they would still maintain what we experience is at best merely a temporary aberration, or at worst a near-universal illusion of the increasing obsoleteness of class. For them it is simply a fact that in the long run, class will return to structure the political sphere, even if only in the final analysis. However, it is far more realistic to claim that if class is still to be hegemonic, it must be transformed to mean something quite different from how it was conceived in nineteenth-century Marxism. Part of this transformation could be seen as informed by the risk society thesis.

Risk places a whole new reflection on class politics, one that provides a rather different arrangement of 'alienations'. Whereas in industrial society, alienation involved a gradual disconnection between workers and the product of their labour, the labour process, their fellow workers and, ultimately, themselves, alienation in the risk society has completed this movement of separation by disconnecting the human being's own existential moment from his or her being-in-the-world. That is to say, the human concern for being-in-the-world has been displaced by risk itself. Similar to Giddens (1991, 1994), Beck refers to this process in terms of 'individualization'. Individualization does not simply mean a growing sense of isolation, i.e. alienation from society (individuation), but a process of disembedding of the human being from his or her socio-cultural milieu (a process also referred to as detraditionalization) and a subsequent re-embedding into a world ordered and revealed by technology, that is, a world of 'know how' that is the exclusive property of expert systems and technologies of mediation. In this sense individualization is also a form of societalization (Beck, 1992: 90), a movement from *Gemeinschaft* to *Gesellschaft* (Tönnies, 1887/1957).

In western societies, it is generally believed that 'modernity' provides a most general and abstract framework of understanding moral issues. Authors as divergent as Bauman (1990), Giddens (1990, 1991), Lyotard (1979) and Vattimo (1997) have all argued that modernization entails a process by which judgements of right and wrong are increasingly divorced from traditional religious discourses and narratives, and developed in more secular, mundane and individual modalities. Whereas we should at least question the certainty with which traditional moral discourses have been dismissed, as if this process simply mirrors the collapse of theodicity in terms of understanding 'truth' (as validity), it cannot be denied that, nowadays, the percentage of people in western societies who look towards religious institutions and established religious dogmas for moral guidance is much smaller than 100 years ago.

Like Tönnies, Giddens (1985, 1990) sees the decline of traditional, face-to-face oral communities in the nineteenth century due to a combination of forces: industrialization, urbanization, the development of the nation-state (and its monopoly of legitimate means of violence and taxation) and of course technologies of communication and transportation that provided the infrastructure of emergent globalization. As a result, human beings are being increasingly uprooted from their traditional, local cultural embedding, entering a new, massive world of anonymous 'citizens' who engage with each other on the basis of more abstract rules of conduct such as contracts. This de-personalization of the *socius* heralds the birth of modern society, and its new political order based in the public sphere.

There is more than a touch of Durkheim, Weber and Parsons (1973) here too. As the *socius* becomes more abstract, sociability becomes more and more organized around externalized, specialist functions, supported by legal-rational instrumental reason. The increased differential and specialized organization of the economy (division of labour) thus finds its match in an increased functional differentiation of the state and its institutions, as the modern bureaucracy emerges and extends itself to all kinds of previously private functions: education, health care, recreation and social welfare.

A final twist in Giddens's story is inspired by R.D. Laing, and has a bit of a Marcusian flavour. In the modern, functionally differentiated, abstract society, people's sense of identity is being threatened, as traditional embeddings are uprooted. This generates a sense of ontological insecurity: people do not know who they are and they do not know why they are here. Because of the increased complexity of social life, traditional moral guidelines no longer suffice. People are thrown back onto themselves; they know they exist but have lost a sense of meaning or purpose of living. Hence, they have to constitute their own moral guidelines to find ontological security. Giddens, however, turns away from the pessimism of Marcuse (and Laing) by stressing that this sense of alienation is being compensated by increased reliance on expert systems, forms of counselling, care and guidance based on scientific knowledge aimed to help the individual attain a sense of happiness and meaningful existence.

Beck argues that the process of individualization is furthered by the risk society as it produces a new language of being in terms of risk. It engenders a sense of 'existential insecurity'. Individualized beings are thrown back onto themselves; they must decide whether to use a condom, wear a seatbelt, touch that stranger, eat beef or not. Media technologies make an enormous amount of information available about possible and definite risks, but individuals must negotiate between them. The information is produced by science, governance but also commerce which thrives on the transformation of risks into opportunities.

A good everyday example is domestic cleanliness. After the Second World War, the spreading of mass-consumption society coincided with the arrival of electrical domestic appliances such as vacuum cleaners, washing machines, refrigerators, spin-dryers, irons, toasters, coffee-makers, television, hi-fi and, more recently, the microwave, the PC and the digital telephone. The vast majority of these appliances are now a standard part of any modern western household. Indeed, the modern household is now to a large extent ordered *and* revealed by these technologies. With these technologies, the household has been transformed from a labour-intensive 18 hours-a-day production unit, to a more technologically intensive leisure and consumption unit. It is therefore remarkable that women in particular still work many hours a day to 'produce' a comfortable and clean household environment. This is because with the new technologies, the standards of hygiene and speed of turnover also increased; i.e. the technologies have revealed a world of housekeeping which requires more attention to a greater number of details.

This transformation of an ethos of hygiene started perhaps a century before the arrival of domestic appliances, with the sanitation movements which were often closely associated with a rhetoric of national (military, economic and cultural) strength (Coley, 1989; Latour, 1988; Roderick, 1997). However, whereas the early sanitarians associated with technoscience, in particular those involved in engineering 'germ control' (vaccination, pasteurization), the mass-consumer hygiene drive was mainly generated by a technological culture in the form of mass education and mass entertainment.

Every technology brings with it its own sets of norms and standards. Technological culture is inhabited; it induces a specific habitus. The habitus associated with many of the domestic appliances has been that of increased 'health' and 'hygiene'. In particular since the 1960s, there has been growing awareness that cancer and cardiovascular diseases, the main causes of death in affluent societies, are related to practices of consumption. Moreover, developments in medical technology, in particular the discovery of antibiotics, enabled people to reconceptualize 'infectious disease' as something they can partly control and even conquer (Van Loon, 1997a). Already since the Victorian household sanitation movements of the late nineteenth century there was a growing sense of the dangers of invisible creatures ('germs') as the cause of many diseases. However, with new medication alongside a growing range of chemical disinfectants used in soaps and other cleaning agents, the warfare against germs began to focus on the household itself. More recently, it has become clear that many bacteria have become resistant to antibiotics, and causes of this have been linked to the extensive use of antibiotics in medicine and also in the food industry (Cannon, 1995; Garrett, 1994). Frequent use of disinfectants have sterilized our environments and inhibited the development of immunity against bacterial infections. This has given rise to deadly forms of bacterial meningitis, as well as the emergence of new highly virulent forms of streptococcus, pneumonia, ear infections, etc. Risks of infectious disease are thus coupled with technologies of domestic sanitation. First, they were part of the risk management, but subsequently, they emerged as risk factors themselves.

However, not only in the spheres of domestic consumption was there a growing awareness of the relationship between risk and technology. In industrial production too it became apparent that technological transformations revealed new forms of risk. In the petrochemical (W. Marx, 1971) and nuclear industries (Welsh, 2000), strange diseases emerged among the employees and also among those living in or near the industrial sites. Today, risks of chemical pollution and radiation are part and parcel of our everyday existence. We have all been exposed to higher levels of pollution than our grandparents. However, disasters such as those at Bhopal in India and Chernobyl in Ukraine reveal that these risks can be extreme and imply massive deaths. Together with more general ecological disasters such as the depletion of the ozone layer, soil erosion, climate change and so on, we can see that modern technological culture certainly engenders a risk habitus.

In short, we can clearly see that technologies and risks are intricately related. More importantly, it becomes obvious that technology contributes not only to the production of hazards, but also to our understanding of those hazards as risks imply a form of agency. They are produced in specific forms of social and economic organization but always require a symbolic form to come into being. Risks are always threatening to take place, they never take place (as disasters do). They are events-in-becoming (Beck, 2000; Van Loon, 2000a). Without symbolic forms, risks are nothing.

Hence, we need to understand the crucial double role of technoscience in the production of risk: first, it generates 'risks' by the transformation of nature (for example genetics); second, it grants us insight into risks. There is a third factor here: technoscience further reproduces risks by granting us insight into risks, because it presents them within the specific technological constellation that is made available. Technology both reveals and orders our being in the world. Like the domestic technology that reveals new 'problems' of cleanliness and hygiene to the '1950s housewife', or genetic technology that reveals new and previously unknown genetic defects, these technologies also order and reveal these revelations in particular ways of seeing and understanding. This ordering and revealing produces more investment of energy and resources into the further improvement of the technology.

Using Beck's insights, we can now argue that individualization is a specific form of en-presenting subjectivity, engendered by the risk society's technological culture. As a fundamental concept of modernity, the individual is the centre of all that modern society stands for: it is the essential component of both democracy and markets. Now Beck argues that towards the later stages of industrial modernity, something strange is happening. Increasingly we are confronted with the limits of economic growth; the ecological degradation that accompanied industrial capitalism is now producing so many negative side-effects (mainly in the form of chronic as well as infectious diseases), that humans have begun to seriously question the values of 'progress' and 'rationality'. Science and technology are being approached with far more suspicion than before.

Although there have always been major disagreements between scientists, the belief has persisted that given the right resources and in time, the truth can be discovered. Today, disagreements between scientists appear to be more irresolvable. Scientific authority is waning. Its impotence became particularly clear during the BSE-crisis (Van Loon, 2000a). The effect of this is a further intensification and shifting of the burden of responsibility for decision-making to the level of the individual. Do we eat beef or not? What are beef products anyway? In other words, we are forced to think about things we normally do out of habit; we have to become more reflexive, just as the '1950s housewife' had to become more reflexive about cleanliness with the arrival of domestic technologies. This reflexivity comes first as an upsetting of the habitus, in the form of self-confrontation; but in order to work, it has to become part of that same but now transformed habitus. Reflexivity is thus not a permanent state of conscious being, but an existential moment that can only have an effect through a repetition of its forgetting and its emerging.

It is also clear that to operate in this world of individualized responsibility consumers require not only far more information but also the skills to interpret them. The catch is that this information is itself fed by the same technoscience that generates the risks. In other words, 'conscious consumption' is itself enframed by the technology that revealed its risks. This paradoxical phenomenon repeats itself in discussions over genetically engineered food, over waste and packaging, over transportation, over diets, etc.

In modern society, technology always had a strong pedagogical function. Technological innovations were frequently accompanied by educational strategies for their dissemination in wider society. Magazines such as *Good Housekeeping* have been invaluable in bringing housewives up-to-speed in terms of domestic technologies and the risks they order and reveal. Today, such magazines still thrive on the vocation of conscious consumption but it could be argued that their consciousness is no longer in unequivocal support of science and technology. Recent outbreaks of BSE and foot and mouth disease in the UK have undermined consumer confidence in the food industry, despite scientific guarantees about food safety. Likewise, despite strong scientific campaigns for genetic engineering, distrust of genetically engineered products is widespread. Technoscience responds with calls to increase the public understanding of science (Rose, 2000; Irwin, 2000), but this response fails to acknowledge that technoscience is now increasingly seen as part of the problem.

Individualization is thus intrinsically linked to a technological culture which orders and reveals our being in the world in terms of risk. Neither Beck nor Giddens would deny that this did not happen with industrial modernity; quite the contrary for Beck, resonating many of the classical sociologists, scarcity and need were central forces in generating individualization in the first place. However, the main difference between industrial and risk societies is that whereas in the former, institutions and technological systems such as the welfare state, provided a relatively stable framework in which these risks could be assessed and managed, in the latter, such institutions persistently fail to transform risks into securities. Risk containment has itself become contaminated.

This sense of individualization of needs through risk is not simply a matter of addition (i.e. both scarcity and risk) to industrial society, but one of translation: risks turn scarcity into something else, a distorted mirror image of itself. The language of risk is itself contagious. It induces everything into its own properties. As a result of this, it makes a mockery out of the positions held in the traditional class struggle for which scarcity of work has become the risk of unemployment. Indeed, class struggle today for example involves a wider set of risks and evolves much more around health-and-safety issues. In the face of the catastrophic potential of nuclear or genetic disaster, class struggle based on scarcity simply ceases to be all embracing. Risks undermine the systemic integrity of industrial society and its associated social formations: social class, the (patriarchal) nation-state (bureaucracy) and its disciplinary apparatuses. It reorganizes inequalities on the basis of risk relations. More often than not, however, these inequalities will take place along familiar lines as those regulated by scarcity: poverty, social exclusion, gender, ethnicity, disability and so on, but – one could argue – with increased volatility and internal differentiation. That is, because of the close proximities between risks and opportunities, the discrepancies between collective and individual rates of success and failure will increase, making it more difficult for an organized hegemonic strategy based on equivalences between, say, women's movements, civil rights movements, anti-racism movements, etc.

However, Beck stresses that individualization cannot be equated with just disembedding. Similar to Giddens after him, he associates it also with a second movement – towards re-embedding. That is to say, Beck sees individualization as reinforcing new dependencies between institutions and social systems. Existential insecurity engenders anxiety and distress. In order to cope, the individual turns to institutions and expert systems such as counselling and psychotherapy. However, most of the coping is already arranged and enabled by (in terms of both Foucault and Parsons, see Kroker and Cook, 1988) 'normalizing' expert systems such as family planning, medical care, education, social services, prisons and other disciplinary systems. All these are examples of how previously communal social systems (e.g. the family, the tribe) have been eroded and transformed into state apparatuses.

Individualization is then used to describe this dual process of (a) uprooting from traditional collectivist culture and (b) reorientation on a personalized and biographically assembled grid of ontological security. However, it would be quite wrong to suggest, as some have done (e.g. Lupton, 1999), that the theses of Beck and Giddens on individualization are the same. For Giddens (1991), individualization is a social psychological adaptation of human beings to living in a *Gesellschaft*. It is fundamentally linked to a process of detraditionalization. In this sense it is emancipatory, even liberating, as the individual becomes more free to address his or her needs in ways he or she prefers. Moral obligations thus become restricted to those imposed by individuals themselves. This sense of individualization fits in with a liberal ideology of personal freedom and choice. This ideology, which emerged from a collusion between modernity and the enlightenment, has been culturally dominant since the nineteenth century, but one could argue has now nearly attained full hegemonic status as very little can be successfully presented as a viable alternative to it.

This, however, is not the only way in which one could interpret individualization. Beck differs from Giddens in that his insistence on risk enables us to see a darker side, one that reminds us of the pessimism of Marcuse and Adorno without becoming entangled in their tragically utopian idealism. For Beck, detraditionalization is not the only force behind individualization, risk is a different kind of ontological insecurity than merely a side-effect of detraditionalization. This is not just the sense of risk as 'manufactured uncertainty' which Giddens recognizes, but a sense of risk engendered by overdetermination and complexity. For Beck, expert systems may offer themselves as solutions by developing expert knowledge about particular risks, their causes and the best ways of handling them, they are inherently contributing to an exacerbation of risks, because no system can effectively control the complexity; there is no form of knowledge that surpasses the indeterminability of risks (Adam and Van Loon, 2000).

For Beck, individualization is a default outcome of a *failure* of expert systems to manage risks; neither science, governance, media, commerce, law nor even the military are able to provide sufficient closures of risks to enable people to place their trust in these institutions. As a consequence, people are thrown back onto themselves, they are alienated from traditional communal systems but have nothing else instead. Hence, responsibilities for decisions which entail risks become personal and private: to eat or not to eat hamburgers; to have or not to have sex; to have or not to have a pre-emptive mastectomy because one has been identified as having a specific gene linked to breast cancer (Beck-Gernsheim, 2000; Rose, 2000). There is nothing free about this individualization as it is forced upon us by default. The choices one has to make are not free choices, but informed by incomplete, often dubious, information from not always trustworthy sources.

These two accounts of individualization thus result in quite different political-ideological tenets. As already mentioned, the Giddensian line embraces liberal individualism within a framework of institutional arrangements; Beck's line is more critical: individualization is paired with anxiety and a continued ontological insecurity, manifesting itself in 'risk cultures' that inhabit the risk society. In the first line, morality follows individualization and becomes 'a personal issue', in the second line, morality is undermined by individualization that is driving a culture of apathy and indifference, sporadically interrupted by manic outbreaks of panic and anger.

Beck's more critical stance exposes a problematic assumption in the work of Giddens. It seems to suggest that modern expert systems actually manage to provide a new existential security. Indeed, as Foucault's (1977a, 1980) work so clearly demonstrates, these aforementioned disciplinary institutions generate particular forms of modern being that are easy to recognize: the patient, the student, the client, the prisoner, the private, the employee, etc. These are presented as being distinct and discrete 'identities'. However, it makes no sense to suggest that these institutional identities have ever been formed without contestation and are themselves inherently stable. There is a wealth of evidence to the contrary. Despite all the disciplinary work, the docile bodies thus produced do not correspond with docile subjectivities. Moreover, despite the regimes of identification enforced by these institutions, there is no evidence that these docile bodies inhabit less existential insecurity than those 'cases' outside of the domain of institutional modernity (Rhodes, 1998).

This should not come as a surprise. If one follows the earlier writings of Tönnies (1887/1957) or Durkheim (1984), it becomes nowhere clear that the transition to modern society should be expected as a smooth one. Both predicted that new social formations will be riddled with instability and fragmentation.

What Giddens (1991) describes as the necessary response to existential insecurity in a detraditionalized society is in fact a tragic failure. Trust in expert systems immediately raises the spectres of suspicion and scepticism. Without membership to 'primordial' collective units such as the tribe or the family, the modern individual seems forever lost in a myriad of significations and myths which can never be secured or anchored in trust, because this trust is itself virtual. Surely, Beck and Giddens do not claim that modern institutions are as functional in practice as they are in theory, but Beck has a clear advantage over Giddens in that his central focus on risk attunes him immediately to the failures of modern institutions and the individual's dependency on them. In this sense, the re-embedding part of individualization is nothing as effective (and affective) as its disembedding. This is similar to what Habermas referred to as *privatism* (Beck, 1992: 98). Lodziak (1995) defines privatism as follows:

> the growing disinterest in politics, reflected in the declining membership of traditional political parties and trade unions, together with an increasingly wide-spread tendency for people to place their energies into private solutions to the problems they face.
>
> (Lodziak, 1995: 75)

For Lodziak (1995: 75), this suggests 'a *loss of faith* in collective and political solutions' (italics mine). This loss of faith itself suggests a continuation of the secularization process that characterizes modernity as a whole. It is simply extended from abandoning faith in God to losing faith in 'man' – as a collective being – 'himself'.

A crucial element in understanding the ambivalence of re-embedding in Beck's individualization thesis is 'reflexivity'. Reflexivity is that which 'returns' from a particular experience that was not anticipated in the encounter itself. It is an essential part of any learning process as it involves feedback from one's environment on the effects and outcome's of one's actions. The reflex, however, is also a coping device. It may include a reflection, but is not necessarily limited to cognitive processes (Lash, 1994).

There is a direct link between alienation and reflexivity. Alienation creates a distance, a disparity between the existential and hermeneutic subject. For Hegelians and the young Marx, this alienation is above all a self-externalization, like the Lacanian mirror-stage in which the infant becomes a subject (Lacan, 1977: 1–7). The self is doubled in the articulation of the 'I' and the reflection or referent of that 'I' (e.g. 'I am me'). It becomes objectified in an image, in a figure of speech, in the grammatical subject. This alienation is perhaps best described in terms of what Scott Lash (2000), liberally appropriating Kant, has referred to as 'reflectivity'. The alienation that inaugurates reflectivity is tragic, that is, it inaugurates a strangeness that drives a quest for reunification (e.g. trying to 'be oneself' is always a failure for its achievement can only be a deception). This is the alienation of a migrant trapped in a diaspora, who – at some point after leaving home – discovers that there is no home anymore (Gilroy, 1993; Chambers, 1994).

However, there is a second kind of alienation. This is not a self-externalizing one, but one which turns upon itself on the inside. In Julia Kristeva's (1988) words, it is a 'strangeness within'. Rather than between the existential and hermeneutic subject, it disrupts the integrity of existence itself. Scott Lash (2000) refers to this as 'reflexivity'. In contrast to reflectivity, reflexivity is not primarily operative in terms of cognition or consciousness. That is, it does not appeal to a set of principles on the basis of which the subject determines a judgement. Reflexivity is more primary than that. It *comes before* the subject in the mirror stage, in the form of a lingering radical doubt that the 'I' is already split in a desire for being-there (Copjec, 1989). This is why Beck equates reflexivity with self-confrontation, even self-destruction. Rather than a migrant, the reflexive subject is a nomad who must break up her/his residential state of being in the world in order to be a nomad. Reflexivity does not fix the subject in a mirror image, it disperses subjectivity across a wide range of multiple reflections, i.e. what Haraway (1997) refers to as 'diffractions' (see Chapter 4).[10]

Because reflexivity is not located in consciousness, the self-confrontation it engenders may happen without anyone noticing. In terms of the previous section, reflexivity is the moment in which a bad turns back onto a good, without being annihilated by it. Since these bads increasingly operate at the molecular and submolecular level (atom, gene, virus), their distribution is no longer observable directly by our senses. We resort to technological mediations to visualize, signify and valorize their presence (see Chapter 3). From now on, only a suspense of doubt and suspicion can temporarily unify subject and reflection. Sustaining an integral, individual self in the risk society is indeed a tragic affair.

The communicative logic of risk: autopoietic systems

We have seen that Beck's risk society thesis rests on the assumption that technologies reveal and order particular ways of our being in the world. On this basis it argues that coming to terms with this world increasingly entails a confrontation with the catastrophic potential of risks. This confrontation is a self-confrontation (reflexivity) in which human beings are increasingly thrown back onto themselves (individualization).

The reflexivity produced by a growing disparity between disembedding and re-embedding suggests that in the risk society, social order is gradually being eroded at the cultural level. Indeed, the coupling of the paradoxical nature of the risk society with a new sense of alienation as existential insecurity highlights that for Beck, the cultural dynamic of social forms is central to understanding risk. This cultural dynamic involves not only the pivotal role of sense-making and perception in the actualization of risk, but also what could be called the communicative logic that underscores the very possibility of 'action' of social movements, political parties, institutions, including science and mass media, and corporations. Understanding the full extent of the central role of communication in the risk society thesis requires a more detailed appreciation of the wide-ranging influences of, among others, the works of Luhmann and Habermas, on the development of Beck's thought.

With both Luhmann and Habermas, Beck shares an analysis of modernization as increased complexity and ambivalence. A silent presence of Durkheim can be felt here as this complexity is produced by ever-increasing differentiation. Similar to Luhmann (1990) and unlike (Parson's) Durkheim, however, Beck does not suggest that 'integration' necessarily accompanies the inevitable 'differentiation' engendered by modernization. In other words, his conception of the social stems from a sensitivity to complexity and instability. It pictures social organization as an accomplishment and not a matter of fact.

Central to Luhmann's work (e.g. 1982) is the suggestion that the immense transformation of social order often referred to as 'modernity' can be characterized as the intensification of flows of exchanges (goods, services, people, finance, information) in terms of frequency, velocity and scope. Luhmann's attempt to come to terms with the immanently increasing velocity and pervasiveness of 'flows of exchanges' is called 'system theory'.

The primary characteristic of identifying a system is its difference from its environment. A system is a more or less deliberate set of processing-flows, which interact with their environment in order to modify it. The ability to make this distinction is essential to the formation of any system. For Luhmann (1990), 'system' ranges from the simplest life forms such as that of protozoa to the most complex elaborate human artefacts such as 'society' and 'law'. Systems can come into being only if they acquire the ability to differentiate themselves (create an environment in/of/for themselves). This differentiation is thus in essence a completely internalized and self-referential accomplishment.

The focus in Luhmann's theory is the dynamics of 'communication' within and between systems and between systems and their environment.

> The system of society consists of communications. There are no other elements, no further substance but communications. The society is not built of human bodies and minds. It is simply a network of communication.
>
> (Luhmann, 1990: 100)

Communication, he argues, is a temporal process which results in an ongoing process of differentiation (Luhmann, 1982: 302).

> All in all, a pronounced trend toward greater differentiation and specialization is discernible and hence also a need to institutionalize the arbitrary to an ever-increasing extent. At the same time, the pace of change is gradually accelerating, as generally happens in the course of human development, so that means of overcoming increasing improbabilities in ever faster succession have to be developed out of what is already available, a task that becomes increasingly unrealistic if only on account of the time factor and leads to selection by the criterion of speed.
>
> (Luhmann, 1990: 93)

The emphasis on a system of communication in relation to an ongoing differentiation of society highlights that for Luhmann the constitution of 'the social' is characterized by improbability rather than necessity, diffusion rather than evolution, chaos rather than order and increasing disintegration, not cohesion. Luhmann's world is one in which systems are in a continuous process of differentiation, in which they self-referentially create boundaries between themselves and what they order and reveal as 'their' environment, in order to sustain themselves.

The production of an 'internal environment' as the highest level of systemic development is thus a 'self-re-production' or *autopoiesis*.[11] The maturation of systemic-differentiation is its tendency to become increasingly self-referential and self-generative. Autopoiesis inaugurates a break with the originating system in which 'the medium' (organism, organization, system) establishes a full closure by enclosing and incorporating its own environment (Zolo, 1991: 64). Autopoiesis takes place when a medium

> ihre Systemkomponenten selbstproduziert: sonder kommunikationen als Elemente, selbstgenerierte Erwartungen als Structuren, eigenständige Prozesse, thematisch definierte Grenzen, selbstkonstruierte System-umwelten, selbstdefinierte Identität. Diese selbstreferentiell konstruierten Systemkomponenten sind ihrerseits über einen Hyperzyklus zu einander verknupft.
>
> (Teubner, 1989: 87)

> produces components of its system by itself: especially communications as elements, self-generated expectations as structures, self-standing processes, thematically defined limits, self-constitutive systemic environments, self-

defined identity. These self-referentially constructed system-components are for their part connected with each other by means of a loop.

It is thus the investment of an increasing significance of material embodiment of the medium itself. McLuhan's famous expression 'the medium is the massage [sic]' (McLuhan and Fiore, 1967: 157) simply means that since all media are extensions of 'man' the (technological) environment that human beings create becomes the medium for defining their role in it. *Mediation* is thus not a passive transportation of information from A to B, but itself an engine of meaning creation. The autopoietic 'medium' engages in self-observation and self-description. Indeed, 'the content of any medium is always another medium' (McLuhan, 1964: 23). Self-referential systems are sovereign with respect to the constitution of identities and differences (Luhmann, 1990: 3).

Autopoiesis has considerable implications for the way in which the communication process can be understood. Through the notion of autopoiesis, system theory shifts the conception of, for example, mass media from functional interdependence to functional independence. As such, the notion of 'function' becomes devoid of any explanatory power and should be replaced with 'operation'. 'Society' becomes nothing more than a communication system: 'Die Gesellschaft besteht aus Kommunikationen die die Eigenschaft haben, andere Kommunikationen zu produzieren' (Society consists of communications that have the characteristic of producing other communications) (Teubner, 1989: 87). Functional evolution thus becomes erratic dispersion, a fractal whose outcome is as unpredictable as the process by which it takes form (Luhmann, 1990: 182).

The autopoietic communication process in system theory consists of three moments of selection: information, utterance and understanding. As Lyotard (1991) has argued, information is the hybrid of technology and energy (matter). The utterance is the performance (see p. 12) whereas understanding is the transformative response, a temporary projection of the completion of the communicative process (identification). Luhmann insists that these three selections need to be synthesized in self-referentiality.

> Without a synthesis of three selections – information, utterance and understanding – there would be no communication but simply perception. By this synthesis, the system is forced into looking for possibilities of mediating closure and openness. In other words: communication is an evolutionary [sic] potential for building up systems that are able to maintain closure under the conditions of openness. These systems face the continuing necessity to select meanings that satisfy these constraints. The result is our well known society.
>
> (Luhmann, 1990: 12–13)

In other words, by reconceptualizing sociality as the *result* – instead of the *condition* – of communication and linking this to the inevitability of selection, Luhmann's account of communication or mediation as autopoietic presents an emphasis on 'closure' as a contingent necessity for the realization of order which – because of its arbitrariness – is inherently unstable.

Indeed, if systems become systems because they successfully enclose themselves into their own externalized environment, which is the sole effect of its own internal differentiation, it seems extremely unlikely that different systems will ever be able to communicate. For example, as Gunther Teubner (1989) has argued with reference to legal systems, if the law manages to fully enclose itself into its own self-referential environment, it will only recognize the logic of law as valid. To some degree this is indeed true (as people who have tried to combat the law with common sense constantly rediscover). However, it is also evident that law presides over and interferes with a range of other systems, including the very powerful and largely self-referential system of finance and economics. Teubner argues that systems such as law have the ability to reflexively incorporate the logic of other systems in order to increase its own self-referentiality. This process of *conversion* or *translation* is of extreme importance to the sustainability of any system; not only must it be able to differentiate itself from its own environment, but also it must incorporate other systems and their environments and either induce or scan in their logic of differentiation and self-referentiality.

Confronting subpolitics

Although he shares with Luhmann a fascination for the complexity of open systems, unlike Luhmann, Beck does not take autopoiesis as the sole model of systemic development. That is to say, whereas for Luhmann, systems would ultimately resort to self-enclosure in order to cope with and reduce the complexity and enable the continuation of communications, Beck acknowledges that self-enclosures (for example, individualization as disembedding) are one strategy among others (for example, individualization as re-embedding). Hence, like Teubner, Beck recognizes the possibility of conversion between different systems. As the example of domestic hygiene given earlier in this chapter indeed shows, the systems of technoscience, governance, media and commerce have been highly effective in conversing a wide range of risks into seemingly tangible practical operations of everyday life.

Consequently, Luhmann's insistence on the improbability of communication receives less attention in Beck's work. Beck is more concerned with the way in which different systems affect each other, which may mean breakdown as well as synthesis. That is to say, he does not a priori rule out that new systemic configurations between institutions might enable new forms of re-embedding. Moreover, he does not share Luhmann's political quietism.

As his rather critical analysis of individualization (as privatism) already suggests, Beck's own political ideals are much closer to those of Habermas. For Habermas, the public sphere is characterized by 'reasoned critical discourse'; it is generated by both the emergence of a liberal bourgeois class of industrialists (opposing the privileges of the feudal aristocracy) and the subsequent (working-class) struggles against the most visible negative consequences of unbridled capitalism. The role of religion and philanthropy should not be overlooked here, as much of what subsequently became known as the welfare state (which is often seen as one crucial institutional realm of the public sphere) emerged directly or indirectly from such privately organized charities (De Regt, 1984; Jones, 1985).

Habermas (1989), however, is also very critical of the sort of institutions that have emerged out of these struggles; he sees them as being dominated by, on the one hand, a sort of culture of commerce – public relations (PR), advertisement, marketing – and on the other hand a rather instrumental understanding of rationality (in terms of means to obtain ends, without due concern for the procedures by which such means and ends are being linked). Habermas's political philosophy could be seen as an attempt to reclaim the public sphere as a genuine space for actualizing democracy, on the basis of reasoned communicative action (also see, for example, Dahlgren, 1995; Thompson, 1984, 1990, 1995). In his most famous work, *Theory of Communicative Action*, Habermas (1981) sets out a procedural rationality for reconstituting the public sphere through a revitalization of political and public dialogue and participation. Habermas developed this on the basis of a distinction between what he calls system (capitalism, commerce, bureaucracy) and 'life world' (everyday realities and interactions). He locates 'instrumental rationality' in the first category, and communicative action in the second. His thesis is that life world has been increasingly colonized by the system and, as a consequence, has lost much of its practical ethics and morality that governed interpersonal interaction. The way to stop this, he argued, would be to bring back a communicative rationality, that is a rationality of engagement based on a dialogue, a willingness to deliberate a range of alternatives and decide, through consensus, on the most rational/reasonable of the options, ideally in the absence of domination and prejudice.

Like Habermas, Beck has a maintained commitment to a sense of modernity as emancipation, and more specifically, a faith in reasoned deliberation (communicative action) as the main political means by which this emancipation is to be realized in his most recent work on cosmopolitan democracy (Beck, 1998). Like Habermas's thesis of the transformations of the public sphere, the risk society thesis also depicts an erosion of traditional politics, at least in terms of their relative openness to democratic participation. Technocracy, bureaucracy and commodification (instrumental rationality) are identified by both as archenemies of democracy; and they are ubiquitous in the risk society.

Although Beck's politics are more eclectic and pragmatic than the philosophically based ideas of Habermas, the latter's thought has clearly influenced that of Beck. His emphasis on 'another modernity' which he – with Giddens and Lash – terms 'reflexive modernity' (Beck et al., 1994), must be seen in this light. What is at stake here is to repoliticize social theory in terms of enabling a critical self-confrontation that is already offered to us in terms of risk, and extending these to reflect on the way in which the institutions of modern society have failed to address risk-related issues properly, because they are still captivated by categories and concepts from a previous era.

Inspired by Habermas, quite a few commentators have focused their attention on mass media as the emergent 'public sphere' (Dahlgren, 1995; Stevenson, 1995; Thompson, 1995; Webster, 1995). John Thompson (1995), for example, argues for a reinvention of publicness. He relates this to a notion of publicness that is not so much derived from the separation of public and private as from the separation of visibility and invisibility. He questions the usefulness of Habermas's implicit preference for co-presence in his 'ideal speech community', and wants to develop a new notion of public sphere, based on a more positive acknowledgement of the transformative role of the mass media. It allows a notion of public sphere that exists beyond the physicality of space. For him, such an emphasis on mediated public sphere could lead to a politicization of the everyday.

Beck follows Habermas quite closely in his call for a revitalization of a deliberative democracy – a democracy of reasoned engagement. For him, 'sub-politics' are inherently based on an instrumental rationality which, when coupled with the privatist re-embedding of individualization, lead to a danger-ous depoliticization of public life and a rather apathetic, if not fatalistic, response to emergent risks.

Reflexivity is for Beck the central concept in understanding the communi-cative logic of the risk society, or better, it is that which will enable us to engage the risk society more positively in pursuit of an alternative modernity. Reflexivity is an effective concept in a critique of instrumental rationality, which – because simply motivated by a desire to maximize effectivity and effi-ciency – does not engender affect and cannot develop a concern for being, which is necessarily a priority in the risk society. Whereas instrumental ration-ality may indeed still adhere to some sense of reflection in terms of means and ends, reflexivity is necessarily unfocused and dispersed because it engages in an unconditional concern for being (which often inaugurates existential insecur-ity). Indeed, reflexivity is essentially connected to scepticism and doubt, which are advocated by Beck not in an ethos of despondency but as disclosures of other possible fragmented futures. Indeed, there are traces of an accomplished nihi-lism in Beck's later works. In this sense, it is wrong to describe the risk society either in terms of utopianism or nihilism. The right word to use here is *ambiva-lence*.

Indeed, the key question of governmentality in the risk society is how to handle ambivalence. Ambivalence is risk. What makes the production and distribution of 'bads' so potent in the contemporary world is the impossibility to evade their implications. Systemic closures such as those generally offered by science in the form of 'expertise', governments in the form of 'legislation' and media in the form of 'moral panic' are no longer an option as we are all implied in this worldwide web of risk technologies. That is to say, reflexivity disassem-bles autopoiesis and reassembles communication flows into hybrid systems (Van Loon, 2000b). Closures offered by expertise, legislation and moral panics are not met with trust in the systems that produced them. The technical fix can no longer be performed.

Risks are infinitely reproducible, for they reproduce themselves along with the decisions and the viewpoints with which one can and must assess decisions in a pluralistic society. For example, how are risks of enterprises, jobs, health, and the environment (which in turn break down into global and local, or major and minor risks) to be related to one another, compared and put in hierarchical order?

(Beck, 1994: 9)

Hence risks force social systems to manage an increasing complexity. At the same time, this handling of complexity exacerbates risks. In response, social systems turn to abstractions, to models, which are, as Latour (1987) has so effectively shown, always of a relatively similar size (about the size of a table top). Reflexivity is called upon to compensate for what remains in excess of the models. It invokes the risks of abstraction with which all risk-assessment models must confront themselves. Therefore, another paradox of managing risk complexity is that with the increasing need for certainty, strategies are devised that – alongside certain kinds of certainty – also produced new uncertainties. Moreover, with the increasing need for decidability, forced upon social systems by their own logic of governmentality, come decision-making strategies that actually engender more undecidability (Adam and Van Loon, 2000; Clark, 1998; Van Loon and Sabelis, 1997). More uncertainty demands more knowledge, more knowledge increases the complexity, more complexity demands more abstraction, more abstraction increases uncertainty. Exactly the same goes for undecidability. In other words, risk management generates more risks.

The systems of risk management consist of science and technology, governance, media and commerce law and the military. In the following chapters, we shall discuss in greater detail how we can understand how these systems work. It suffices here to place them in the wider context of Beck's paradoxical arguments. The main question here is whether Beck suggests that the multiplication of risks is a dead end, or whether he believes there is an escape from the negative dialectic of risk and risk aversion and the entropy of ambivalence.

As can be expected, he is undecided but does suggest that the question of how to confront the risks of risk management will be a fundamental one which will decide on the sustainability of the future of modern society. In Luhmann's terms, the main issue is whether society will remain characterized by the natural response to the improbability of communication, i.e. autopoiesis, or by a radical embrace of reflexivity. Autopoiesis is highly present in attempts by, for example, medical science to insulate itself from legal accountability by setting up its own systems for self-regulation, discipline and enforcement of professional codes of conduct. Traditional notions of expertise are fed by such autopoietic desires. They enable a closure of debate by inhibiting and preventing scrutiny from any other system. Consequently, they cannot be held accountable and do not enter the public sphere.

Modern institutions tend to opt for autopoietic strategies: technoscience and law through the exclusive authorization of expertise; and governance and the military through secrecy. However, there also are strategies that operate not by restricted self-enclosure, but by an expansive conversion. Media, for example, operate through the (self-valorizing) freedom of speech (which significantly does not extend to a critical self-confrontation of its own practices). Most importantly, perhaps, commerce operates through 'free' market exchanges. A hallmark of these two strategies is that, similar to the restrictive forces of expertise and secrecy, they do not enable correspondence with other systems. Free speech is a strong attempt by media systems to translate every operation, every phenomenon, every event into a mediated reality, organized on the basis of the logic of media (e.g. 'two sides of the argument' which excludes the 'third side' – that of the journalist enframing both). Most media, however, cannot even claim to endorse 'free speech', as their operations are actually already induced by a wholly different logic – that of 'free markets' (Berry, 2000). Here, the product of mediation is nothing but a commodity, aimed to be sold profitably on ever-expanding markets. Indeed, the logic of commerce, which in capitalist societies is inevitably bound to the logic of capital, reduces all operations, events and phenomena to financial transactions. It valorizes all flows in terms of capital, i.e. exchange value, thereby disabling, for example, moral judgements on the basis of a common well-being that cannot be expressed in economic terms. For such a system, risks are interesting only if they can be transformed into opportunities.

The politics of industrial modernity have always favoured autopoietic strategies. Autopoiesis is a highly effective way of dealing with complexity without having to deal with any other form of reflexivity than what it generates itself. By isolating only those risk factors it can recognize, a system of unreflexive expertise however, opens itself to new risks emerging from that which remains beyond.

The need for such an integrated approach has been signalled by Beck in terms of reflexive modernization. The primary political unit of this alternative is the individual. Dispersed by reflexivity across a wide range of reflections, the individual cannot be contained exclusively by a single institutional domain. That is to say, for Beck, the promise of individualization as a politically progressive force lies in its disruptive potential *vis-à-vis* the sovereignty of institutional domains. A key aspect for this is the liberation of politics from governance (Beck, 1994: 17). The main reason for this could be seen in terms of autopoiesis: governance has depoliticized the political because what is considered to be political in an industrial society (e.g. 'class struggle'), no longer has any bearing on any system outside governance itself. Hence the title of a more recent work *The Reinvention of Politics* (Beck, 1997) in which he explores possibilities for opening up various self-enclosed systems to political intervention. A key element here is that for Beck, finding political openings is as important outside the domain of governmentality, in science, media and commerce, three systems which engage in 'subpolitics'. Subpolitics are deeply informed by autopoiesis, ways of closing politics into matters of fact.

The reinvention of politics, most strongly advocated in Beck's (1998) recent work on cosmopolitan democracy, necessitates a reversal of autopoiesis, an opening of intrasystemic closures. Reflexivity is exactly that: it disperses the unity of subject and reflection across a range of media, technologies, operations, significations, valorizations. The political challenge of the risk society lies in the systems being able to act upon each other without complete conversion. That is to say, politics that enable communications between different information flows without reducing them to the logic of one system only. For Luhmann, this is impossible; hence his political quietism. For Beck and Habermas, there is no alternative.

Perhaps they engage too much in wishful thinking. However, it is evident that our culture has responded, often in unexpected ways. Under the mushroom of an emergent apocalyptic culture (Doel and Clarke, 1997), new social movements have arisen to politicize what was once the sacred domain of expertise, secrecy, free speech or free markets. Even within those systems, we can observe gradual changes that favour reflexivity. For example, science and its technologies of visualization have fundamentally transformed the 'see-no-evil/hear-no-evil' principle that accompanied the focus of the visible and quantifiable aspects of risks and dangers commonly associated with industrial production. 'Leave it to the experts' is no longer an acceptable slogan, similar to 'Trust me I'm a doctor'. They have become standard jokes in Hollywood films. As invisibility is no longer an excuse for non-decision and non-action, the full implications of the catastrophic potential of industrial production are increasingly becoming part of everyone's being-in-the-world (Adam and Van Loon, 2000). This catastrophic potential is engendered by the indeterminable character of risks and hazards, which has subsequently eroded the politics of security of the finance and insurance complex upon which so much of contemporary capitalism depends.

Reason without faith

Whereas it would be unfair to read the risk society thesis as either a venture in social or technological determinism, there is a danger to either overinflate the role of a collective consciousness, here primarily defined as 'risk awareness' or – inversely – overinflate the political-economic infrastructure of social changes. Beck has been accused by some of his critics as either being too pessimistic (Levitas, 2000), or too optimistic (Rose, 2000), for either ignoring agency or overstating its potential. Beck may have already anticipated such criticism in the turn to a critique of subpolitics in *The Reinvention of Politics*. This, however, may not provide enough antidotes to the suggestion that perceptions and cognitions could change the world on their own (also see Lash, 2000). This problem is generally found among Habermasian political philosophy. The main weakness here is not its inherent 'idealism', but the lack of understanding of the origins of reason itself. By treating it as primarily human, collective consciousness is reduced to a cacophony of individual insights, all competing for attention in the public sphere. This competition is furthermore complicated by the involvement of a wide range of interests, which will not be purified through reasoned deliberation but merely take on more deceptive appearances. It is no wonder that governments are more inclined to give primacy to the voice of a scientific expertise that does not tolerate dissent. Scientists are at a clear advantage here since their alleged truths come directly from expeditions of reality (via 'measurements' and 'instruments of perception').

What Beck perhaps has thus far failed to acknowledge explicitly is what motivates this 'faith in reason'? Whereas deliberative practices undoubtedly enhance the degree and intensity of participative democracy in the public sphere, there remains an unexplored metaphysical dimension to reason, high-lighted by the term faith, which neither Beck, Habermas nor Luhmann have addressed. They seem to reason without faith. Having said that, Beck's obvious appreciation of ambivalence and his boundless creativity in dealing with com-plex issues may actually enable him to address these issues more sensibly and innovatively than many of his contemporaries who remain too much trapped by the aporia of modern social thought. His reflections on scepticism and doubt towards the end of *The Reinvention of Politics* already signal the beginnings of a more philosophical turning to address issues of reflexivity in terms of am-bivalence.

3 Enrolling risks in technocultural practices

Notes on Actor Network Theory

One important lacuna in the work of Beck is the lack of embedding of the risk society thesis in the social logic that is operational in everyday practices of science and technology. His observations are general and rather abstract and do not engage in much detail the inherent ambivalence of technoscience (Clark, 1997a). One way to supplement the risk society thesis and meet this criticism would be to look what has been produced under the label of 'Actor Network Theory' (ANT), in particular the work of Bruno Latour. The notion of 'actor networks' in which humans, technologies and gods are connected in ensembles, or indeed assemblages, is a very useful approach to analysing the way in which particular truths are being established and particular facts being boxed off from further inquiry. The specific notion of enrolment, as a mode of socio-political mobilization, is central to the general argument of this book, namely that technological culture frames risks in particular ways, but cannot contain the contingencies their social and symbolic organization sets into work, as a result of which it destabilizes. Latour's work is particularly useful to generate insights into the way in which risks may constitute a crucial 'agent' within the establishment of technoscience itself.

The enrolment of actors into particular networks is facilitated in terms of 'urgency' and 'necessity' when they are being mediated by a sense of risk. Risk sensibility is fundamental to the mobilization of energy and resources to 'reveal' and 'contain' risks. Risks are what John Law (1995) refers to as 'virtual objects' in the cultivation of technoscientific practices of risk containment. This theoretical analysis suggests that the actor networks that are engaged through risks may be extremely different in make up and form, they will have to engage with three essential practices: attributing insight, attributing meaning and attributing value. These practices however, are never merely instrumental or neutral but 'motivated' by particular political and ethical concerns.

Technoscience in action

> The real is not one thing among others but rather gradients of resistance.
> (Latour, 1988: 159)

Beck's analysis of the risk society as a significant constellation of forces in the formation of reflexive modernization has appealed to a large audience because its macro-sociological logic is well attuned to the dominant political and theoretical discourses circulating in western societies. That is to say, it focuses strongly on the socio-historical, economic and political contexts in which risks have emerged as part of a technological culture. It is less clear however about the philosophical and theoretical grounding of its own reflections on reflexivity, and has swept many of the intricacies of technological culture under the carpet of generalist sociological assumptions.

It could be argued that Beck's understanding of science and technology is overly limited to epistemological concerns about the construction of reality. Although it does not ignore cultural, social, political and economic forces at work alongside ideological and epistemological concerns, Beck's risk society thesis is indeed not particularly revealing in terms of what actually happens in technoscientific practices. It relies on a rather homogenized and sweeping generalist account of science and technology that is more concerned with providing a critique of an image that science constructs for itself, rather than a critique of scientific practices. However, as the crux of the risk society thesis actually evolves around assumptions about the nature of technoscientific practices, a number of scholars have pointed out that very little of the existing literature on science and technology studies has been taken on board by Beck and his followers (Clark, 1997a, 1998; Irwin, 2000; Rose, 2000; Welsh, 2000). Of course, it is always easy to criticize someone for what he or she has not done. The question, however, is whether this omission has any significant consequences for the risk society thesis itself? That is to say, is the risk society thesis compatible with main findings and arguments in science and technology studies? If not, on what aspects could it be revised without disrupting the whole of the thesis? The argument of this chapter shall be that whereas to a large degree the main tenets of the risk society thesis may not be falsified by science and technology studies, if the latter are to be taken seriously, some of the grand claims of the thesis cannot be upheld. This is particularly the case with reference to the implicit historicist (Beck), humanist (Giddens) and teleological (Lash) assumptions that underscore the notion of 'reflexive modernization'. Indeed, we need to get a closer look at technoscience in action to understand how risks are 'produced' and 'reproduced'.

In the risk society thesis, the world of science and technology is largely reduced to its epistemology (following Latour (1998), this will be referred to as Science). In this view, Science is above all a unified monolithic institution, evolving around the standardization of knowledge and its exclusive consolidation in the form of expertise. Although Beck certainly acknowledges the existence of scientific disputes, the dispute itself is seen as antithetical to expertise, i.e. as knowledge not-yet consolidated and standardized. By reducing technoscience to Science, Beck actually confuses two different sets of practices. The discursive practices of legitimation are seen as identical to the practices of 'discursive' production (which also include non-discursive practices). It is as if there is nothing more to say about technoscience than how it wishes to represent itself.[1]

Science and technology studies (STS) are a wide and dispersed field of interdisciplinary and disciplinary approaches to the domains of science and technology (e.g. Berg and Mol, 1998; Bijker and Law, 1992; Mackenzie and Wajcman, 1999). Actor Network Theory is merely one strand of STS, albeit among one of the most influential. It has been a major contribution of ANT, and in particular Bruno Latour, to dismantle the myth that reduces science to Science. This is often confused with an epistemological critique of scientific objectivity, i.e. that scientific practices are not as objective as its dominant epistemology has claimed. The truth of the latter is merely contingent on one's definition of objectivity, itself an epistemological concern. For Latour, something more fundamental is at stake. His work focuses on what techno-scientific practices actually entail. It is the work of science, in laboratories, demonstrations, experiments as well as publications.

The importance of experimentation and laboratory work has been stressed by above all Ian Hacking (1983). Like many others (e.g. Knorr-Cetina, 1981, 1983; Webster, 1991), he focuses on the organized production of science and tech-nology as manufactured within an industry. His thesis is that 'reality' is not constructed in the minds of scientists but becomes knowable through the prac-tical handling of 'objects' (both instruments and 'raw material'). Technoscience is work and studying technoscience has to be understood in similar practical terms. For Hacking, reality itself is not at stake. In this sense, his pragmatic approach, focusing on the way it is being reproduced in scientific practices, is very similar to Beck's.

In contrast, Latour's work is most commonly associated with 'ethnography'. As a result, it tends to bracket off questions of the real-ness of reality, in favour of a stance that is agnostic rather than pragmatic, and is more critically con-cerned with multiple and contested constructions of what we come to know as reality. ANT's main concern is what actors actually *do*. However, although it includes quite a bit of participant observation, it is not entirely reducible to the classical anthropological understanding of the ethnographic approach, not only because Latour places reflexivity at the centre of his work, but also because he considers 'culture' to be more than an exclusively human achievement. Similar to the anthropologist however, Latour uses ethnography as a demonstration, with the aim to let the subjects speak *as if* they speak for themselves.

> To make other forces speak, all we have to do is to *lay them out* before whoever we are talking to. We have to make others believe that they are *deciphering* what the forces are saying rather than listening to what we are saying.
>
> (Latour, 1988: 196, original emphasis)

However, philosophically, Latour departs from the basic assumption of ethnography: that of rendering an account of forces of the real world in terms of a separate logic (of anthropological theory). For him, the very tool of 'anthropological reason' is as much part of the description as the so-called 'object' it is applied to. That is to say, in rendering social practices and objects 'descriptable', Latour sees ethnography as technology – an ordering/revealing of our being in the world.

In a nutshell, Latour's project starts from a thought experiment: what if force (reality) and reason (representation) are not separate? At first, this seems quite similar to Nietzschean and poststructuralist philosophical interventions in the western metaphysical tradition. Of particular importance here is the rejection of the Platonic concept of the Idea as anything but the enforcement of normalization of particular forms of thought as knowledge. Without the regulatory imperative of the Idea, Reason loses its transcendental aspirations and becomes a culturally particular form of asserting power. Everything that stems from reasoning, i.e. morals, truth, values, ethics, etc., becomes an expression of a will to power, which is itself merely a bio/psycho-physiological force.

Indeed, Latour often seems to entertain a rather nihilistic outlook on the nature of thought and being in general. However, a more careful reading of his philosophical position reveals that rather than depending solely on psycho-biopolitics, Latour's nihilism is less one-dimensional. Like Deleuze and Guattari (see Chapter 4), he is keen on addressing forces in terms of multiplicities and associations. This non-reduction – which he refers to as 'the principle of irreducibility' – is the cornerstone of his investigations. The principle of irreducibility posits that nothing is ever real or unreal as such, but everything is tried for its strengths and weaknesses – one could perhaps say risks. Indeed, reality stems from resistance to trials. Knowledge is nothing but realization.

This immediately rings bells with the work of Beck, for whom risks are indeed nothing but 'realizations'. Risks are happenings, not of the bads or catastrophes that they refer to do, but of a 'coming-into-being' of a probability of harm, sometimes indeed in the form of anticipated annihilation. However, whereas in Beck's work it still make sense to separate the reality of risks from their constructed nature (i.e. for Beck there is more to the global ecological crisis than a coming together of mediated, scientific and governmental discourses), Latour's notion of realization as resistance to trials engenders a sense of risk whose actuality is always from the start multiplied yet irreducible to its origination.

Whereas this is all still familiar to students of poststructuralism, the subsequent move by Latour to replace 'forces' by the term 'actants' is perhaps a little more controversial. 'An actant can gain strength only by associating with others. Thus it speaks in their names' (Latour, 1988: 160). Hence, Latour's actants are primarily engaged in 'representations' and the primary motive of representation is to endure.[2] The replacements of force by actant, and enactment by representations, create a highly specific understanding of sociality. In Chapter 4 we shall discuss the kinds of shortcomings this may generate. In this chapter, however, it is worth while to explore in greater detail what kind of sociality is formed by actants and their representations.

Actor networks: the fixation of risks

The basic premise of ANT is that actants acquire strength only in association with others. These associations are nested as orderings within wider patterns or orderings. There is never an order emerging from total disorder, there are only degrees of ordering, that is degrees of resistance. A crucial element, which could have been taken straight from Deleuze's *Difference and Repetition* (1969/1994), is that instead of same and difference we should refer to these associations in terms of translations. Identity and equivalence are merely hard-fought accomplishments of translations that have established stable forms maintained by force (Latour, 1988: 162).

> A force becomes potent only if it *speaks for* others, if it can make those it silenced *speak* when called upon to demonstrate its strength, and if it can force those who challenged it to *confess* that it indeed was saying what its allies would have said.
>
> (Latour, 1988: 197, original emphasis)

Hence, for a force to actualize itself, to become an actant, it must reveal itself by concealing itself as another, a repetition that is also a supplement. A force becomes a force only when joined up in a network.

In *Science in Action*, Latour notes that

> the word network indicates that resources are concentrated in a few places – the knots and the nodes – which are connected with one another – the links and the mesh: these connections transform the scattered resources into a net that may seem to extend everywhere.
>
> (Latour, 1987: 180)

Central to Latour's account of 'science in action' is the way in which particular statements become 'matters of fact'. He refers to the process by which this takes place as 'enrolment' – the gathering and connecting of various sorts of resources: financial, symbolic, human, technological, spatial and so on. Via this extension into a network, the particular yet irreducible force that is thus far merely potential becomes 'matter of fact'. As we know, when a claim becomes a matter of fact, it is very difficult to challenge its existence. Used rhetorically, the expression 'as a matter of fact' functions to a priori fence off and divert dissent from what is being claimed. It is a mode of stabilization. The matter of fact becomes an irreducibility, a black box.

The matter of fact is not as much a matter of ideological imposition or deception, but part of the structure of obviousness that constitutes the network itself. The important difference is that such a 'forgetting to question' is not a trick played on the unconscious, but a simple pragmatic consequence of the costs (not just the financial ones of course) involved in excavating black boxes. Challenging a black box is an extremely dangerous (often career-threatening) affair. The costs of challenging are very high as one risks becoming an object of ridicule and losing irrecoverable amounts of credit, as dissidents to the HIV/AIDS hypothesis such as Peter Duesler and David Rasnick have found out, for example.

For Latour, however, networks are not all-powerful uncontested systemic forces. In contrast and despite the huge concentration of resources, they are still a relatively rare and fragile achievement, as much prone to collapse and disorder as to stabilization and expansion. It is the doubling of power and fragility. Much of the investment of technoscience goes into the recuperation of social order from potential breakdowns and instability. It is fatality that is inscribed in the frail links, indicating that rather than stressing the sedimentation of networks into relationships, networks are mobile, temporary and finite.

Latour's main proposition, that humans, technologies and gods constitute 'actor networks', forces us to take into account the particular functions and operations of science and technology as something that sets into work particular mediated contexts that in turn function as self-referential enclosures of 'reality'. The focus on network-space allows for a more situational-locational specific understanding of risk society and risk culture. It also allows us to extend concerns over identification and embodiment in risk cultures with notions of 'movement', 'speed' and 'intensity' that cannot be analysed at the level of 'individual bodies', but force us to take into account the relational-contextual framework of actor networks.

> The smallest AIDS virus takes you from sex to the unconscious, then to Africa, tissue culture, DNA and San Francisco, but the analysts, thinkers, journalists and decision-makers will slice the delicate network traced by the virus for you into tidy compartments where you will find only science, only economy, only social phenomena, only local news, only sentiment, only sex.
> (Latour, 1993: 2)

HIV is the actant which inaugurates not only a new network, but also simultaneously a series of closures within networks. A reterritorialization of 'risks' takes place with closure. Again, the differences with Beck are not huge, and certainly not insurmountable. However, Latour's uncompromising insistence on 'actants' as organizing forces of realization engenders a more consistent and coherent understanding of the ontology of risk. More importantly, it no longer requires any assumption about the unity between technoscientific epistemologies and practices, which underscores much of Beck's implicit notions of science.

In *Science in Action* Latour's main concern is with the *sociologics* of reason (1987: 205), focuses on 'mapping' as a way of bringing everything into a singular scale on which the event can be 'handled'. Cartography, photography, modelling and simulations allow scientists not only to have a clearer *overview* of the event by making it 'present at hand' (presentation) but also to enrol through translations, new elements that may further stabilize and strengthen the network system (representation). Mapping is a clear act of translation, of transforming an element into a connectable, recognizable, communicable object, an actant of the network. It allows actants to 'act at a distance' (Latour, 1987:

232). This 'scaling down' of the world is thus also a process of intensification of information: the formation of 'nodes' in networks.

Like models, stories enable the formation of nodes in networks. The specific worlds of technoscience are made visible through particular stories for example those told by scientists, politicians and journalists. The starting point for Latour is to trace the paths set out by the stories themselves.

> The fact that we do not know in advance what the world is made up of is not a reason for refusing to make a start, because *other* storytellers seem to know and are constantly defining the actors that surround them – what they want, what causes them, and the ways in which they can be weakened or linked together... The analyst does not need to know more than they... The only task of the analyst is to follow the transformations that the actors convened in the stories are undergoing.
>
> (Latour, 1988: 10, original emphasis)

As with Beck's risk society, ANT thus implies a 'communicative logic', which refers to the process of translation through which sociality is stabilized. For Latour (1999a), making present – presenting, re-presenting and representing (see pp. 53–54) – is primarily about translation. Through translation, sequences of events, of causes and effects, supported by links, come into being.

ANT deploys a very particular concept to refer to the practices of this translation work: *immutable mobiles*. Translation involves a process of making connections between disparate actants, and of making sense of these connections. Immutable mobiles are generated by networks; they are their 'mobile actants' that cannot be silenced (muted), nor transformed (mutated), but instead do all the work of silencing and transforming. Like mathematical concepts, they are 'essential constructions', in the sense that their essence is bound up with the logic of their construction (de Lauretis, 1989) but cannot be modified at will, because their existence is fully integrated with the rules through which they have come into being. A probability calculation, for example, can be derived from population statistics (e.g. epidemiological risk as the quotient of people afflicted from those exposed to a pathogen) or from a mathematical logic (e.g. the chance of a hereditary disease being passed on).[3] However, for each of these calculations, the procedures are fixed and in this sense immutable.

Immutable mobiles enable actants to establish connections with other actants in previously unexplored fields. That is to say, they enhance the number of combinations within a single map or model, and thereby enable the expansion of an actor network on its own terms. Examples given by Latour refer to charts, tables, figures, statistics, maps and indeed models. Immutable mobiles enable 'acting at a distance'.

> All these objects occupy the beginning and the end of a similar accumulation cycle; no matter whether they are far or near, infinitely big or small, infinitely old or young, they all end up at such scale that a few men or

women can dominate them by sight; at a point or another, they all take the shape of a flat surface of paper that can be archived, pinned on a wall and combined with others; they all help to reverse the balance of forces between those who master and those who are mastered.

(Latour, 1987: 227)

Probability calculations are an often-used immutable mobile for the assessment of risks in terms of specific frameworks of action. These are often based on statistical methods and techniques, using large survey samples, census data or information obtained via specific governmental institutions such as education, police or health authorities. They end up in charts supporting specific risk assessments that can be used by people in, for example, government adminis-tration, manufacturing, marketing or insurance agencies.

A well-defined state of affairs is the work of *many forces*. They agree about nothing and associate only via long networks in which they talk endlessly without being able to sum one another up. They intermingle but cannot reach outside themselves to take in what binds them, opposes them, and sums them up. However, despite everything, networks reinforce one another and resist destruction. Solid yet fragile, isolated yet interwoven, smooth yet twisted together, entelechies form strange fabrics.

(Latour, 1988: 199, original emphasis)

However, whereas despite everything Latour's discussion focuses on the strengthening of networks through immutable mobiles – and he certainly matches Beck here in terms of deploying paradoxical reasoning – they may also equally result in the disintegration of specific formations. For example, John Adams (1995) has noted a significant shift during the 1980s and 1990s. What used to be the dominant perspective on risk, advocated by the Royal Society in a report published in 1983 as

the probability that a particular adverse event occurs during a stated period of time, or results from a particular challenge. As a probability in the sense of statistical theory, risk obeys all the formal laws of combining probabil-ities

(cited in Adams, 1995: 8)

has become more contested, unstable and uncertain. What used to be a fixed and certain opposition between objective (scientific) and subjective (perceived) risk has become increasingly eroded of meaning as people act on perceptions, which may not always correspond to what the experts say, but nevertheless 'actualize' reality. The immutable mobiles of risk perception have been under-mined by the risks themselves. How does one actually map the catastrophic potential of a nuclear disaster, or of the continued effects of genetic engineer-ing? An often-heard complaint from scientists about the public's understanding of science is that 'people do not understand statistics' (Macnaghten and Urry, 1998; Wynne, 1996; Yearley, 1991). This could equally be read as that people refuse to understand themselves as a statistic, i.e. they refuse to reduce the meaning of their lives to chance. The immutable mobile of probability calculations is increasingly failing to enrol and stabilize networks around risk, as technoscience is increasingly failing to connect to the other spheres of modern social organization. The public understanding of science could itself be seen as a mode of enrolment, but one whose motivation is overly transparent and perhaps because of this, actually fails to properly engage with the more disparate codes of other systems.

Representations and virtual objects

The concept of immutable mobile enables us to discuss the process of translation in terms of matter flows. That is to say, systems and machines do not simply 'connect'. First, they have to converge in terms of the matter of exchange. Immutable mobiles can be exchanged easily, because systems can still perform their own specific operations on them without transforming the nature of the object being operated upon. Immutable mobiles, however, do transform something: the scales of representation. They enable enlargement (e.g. microscopic operations) or reduction (e.g. aerial photography), in order to engender a sense of 'co-presence'.

Immutable mobiles are thus central in the enactment of presence. They make an object appear, give it meaning and enable it to enter into exchange relationships to establish its relative value. However, whereas initially representation was not problematic within ANT, it has become a focal point of more recent theoretical debates. Latour (1999a) points towards a distinction between presence, re-presentation and representation. Presence refers to the 'object' as such, representation is the production of its intermediation (connecting presence to an absent or distant 'other'), re-presentation is our subsequent re-enactment of the enactment of the object. For Latour the distinction between representation and re-presentation is essential in understanding science as different from religion or art.

Latour (1999b) refers to religion as *re-presentation*, which he distinguishes from science as being mainly concerned with *representation*. Re-presentation is informed by what he calls 'the logic of series', in which faithfulness is a play of repetitions to keep the same message always anew, with the message being one with the messenger (e.g. a priest offering the sacrifice of mass becomes the sacrifice himself). Representations, in contrast, are endowed with the logic of networks, in which faithfulness is the superimposition of inscribed contents whose meaning is a piece of information that does not depend on the messenger or the moral state of the receiver. Whereas re-presentation involves a multiplicity of mediators, representation is the valorization of mediations working towards the annulment of mediators (complete objectivity). Most importantly for this discussion, however, is that re-presentation is always a return to presence and to the present – the here and now – whereas representation is an extension in space-time, which gives direct access to the distant.

Technoscience is primarily concerned with enabling actions at a distance. It is in the business of forming and stabilizing networks. In the modern world, according to Heidegger (1977), it has been so successful in doing that that everything has been turned into a resource for instrumental technological processing. Complete objectivity is achieved when the immutable mobiles fully correspond with the objects they order into presence. That is, the objects themselves become their own immutable mobiles.

For technoscience, representation points towards a relationship between an absence and a presence that is embodied by an object. Risk in technoscience is thus marked by the relationship between the object of risk, and its representation, usually probability multiplied by scope of harm. The problem with risk is that it does not exist without representation. By definition there is no unmediated risk (Beck, 2000; Van Loon, 2000a). Its presence is thus always necessarily deferred. Risk is a potential coming-into-being, a becoming-real. Hence the 'presence' of a risk can never be completely objective but has to be mediated in some form. There always remains a difference between an immutable mobile (e.g. a calculation of probability multiplied by scope of harm) and an (assumed) object (e.g. a hazard).

This difference is perhaps most vividly encapsulated in the concept of 'virtual object' in the sense used by John Law (1995: 284; also see Mol, 1998). Law conceptualizes representation as 'making the knower and making what is known. By creating the distinction between knower and that which is known. And then concealing the connection' (1995: 283).

For Law and Mol, the virtual object inaugurates an epistemological relationship between perception and reflection that is both real (as opposed to possible) and ideal (as opposed to actual).[4] In other words, the virtual object is not a hypothetical entity; it is real in the sense that it engenders reality; yet at the same time it is not 'material' in the empiricist-materialist metaphysical sense of directly observable matter.

Essential to Law and Mol's understanding of the virtual object is the assumption, or even imagination, of a singular entity as both the primordial causal principle of a range of manifestations and 'hidden' beneath the dense texture of practices and discourses of, for instance, management or clinical medicine. For Mol (1998: 150) '[t]his single entity is then projected as a virtual object behind the "aspects" that "surface". This virtual object resides inside the body'. Moreover, the virtual object is being revealed and ordered by these discursive practices and techniques. Its manifestations are being performed as commanding evidence of its alleged presence. Law (1995), furthermore, shows that different discursive practices and techniques may reveal and order different aspects of the virtual objects, or even different virtual objects. What is striking, therefore, is the multiplicity that is inherent in virtual objectification. This multiplicity generates the dense complexity of everyday life practices, including decisions and judgements of what needs to be done.

This also means that despite a surface appearance of unity and rationality, technoscientific objectification entails far more ambivalence and insecurity, in which closures are performed not by following the rules of the scientific method but far more by intuition, symbolic exchange and political association (also see Latour, 1988).

Is this not a return to constructionism and – ultimately – relativism? No, provided that one takes a closer look at what this ordering entails. Ordering has a double meaning. The first refers to appropriating principles for putting things in a 'proper' place. This is the representational notion of ordering that Foucault (1970) identified in terms of the classical disciplines of 'speaking', 'classifying' and 'exchanging'. However, as the modern episteme shifted attention to causation and first principles, 'ordering' became more associated with 'calling forth' or 'engendering'. This is not far removed from the military sense of ordering as commanding. Heidegger (1977) combined both meanings in his notion of technology as a *Gestell* – an ordering or 'enframing' that is also a *herausfordern* – a challenging-revealing – a bringing to the fore or 'en-presenting' (see Chapter 5). Hence, it is not a case of technological determinism by semiotic means. Technology does not fix meaning, objectivity or subjectivity in any specific way; instead it 'performs' a virtual objectification by withdrawing itself – what orders vanishes from what is ordered as if the latter exists independently of the former. This is when representation has successfully transcoded the object in terms of the immutable mobiles that have commanded it into presence. Reality then appears as if it speaks for itself, and thus becomes the most powerful ally in the actor network.

The critical realist notion of 'practical adequacy' (Hacking, 1983) suddenly begets a funny twist. Instead of the image of a critical scientist constructing a concept or technology with which he or she can somehow explain more as an ordering of things (Reality), the scientist is constructing a tool, a weapon, to command reality to perform, that is 'ordering things'. This performativity of method, or technology, owes its effectiveness not to the degree of 'correspondence' between its construction and the real that is outside of it, but to the degree it can order its own information-processing (autopoiesis). This performativity differs between technologies, and these differences matter. Hence, it is not materiality that commands the relative effectiveness of an ordering but the materialization of this ordering commands its own relative effectiveness.

The idea of technology 'en-presenting' objectivity then is neither a form of technological determinism, nor of simple 'constructionism'. Technology as ordering and revealing could indeed be seen as a form of constructing *a* reality, but never Reality as such. Hence, whereas realities are indeed 'constructed', it is not the case that what comes before these is insignificant. Social realities are indeed constructed, but not unconditionally or 'at will'. Instead, reality is what resists trials; it is revealed in technologies of en-presenting that are enabled by the inertia of resistance. This is how we live with technoscience: it provides a temporary ordering, or arrangement, of our world which we may only take for granted temporarily.

However, perhaps because of his aversion to historicism, Latour (1993) refuses to acknowledge the possibility that resistance to particular trials builds up over time (somewhat similar to what Popper referred to as falsification). This is where Beck and Latour part company. This is somewhat unfortunate because ANT is not necessarily to be restricted to 'normal' or 'functional' orders. Indeed, it can be as revealing when applied to what Kuhn (1970), perhaps somewhat clumsily, referred to as 'revolutionary science'. When faced with too many contradictions or problems brought into being by their immutable mobiles, technoscientists do not necessarily have to discard reality in favour of adhering to their method. Instead, they might also reconsider the functionality of the immutable mobiles and construct new ones – to reorder our being in the world accordingly.

Until relatively recently, most risks have been allies to technoscience. For example, bacteria, viruses, chemical toxins, radiation, etc., have all been more rather than less functional in the enrolment of all kinds of resources, usually in service of scientific research. Technoscience was generously supported and – at the same time – its determinations of specific risks were relatively stable and supported by their own demonstrations. However – and this reveals the duplicitous nature of the beast – 'risk' is also itself a specific form of resistance that is revealed in its inertia (projected onto the near future as extended presence). That is to say, risks are not only 'objects' of technoscientific expedition and fixation, but also 'actants' affecting the very nature of technoscientific practice. Risk of failure, for technoscience, is never simply an instrumental phenomenon (negatively linked to 'credit' – cf. Latour and Woolgar, 1979), but inscribed into the very logic of scientific inquiry. It is inscribed in the resistance to trials through which reality is ordered to reveal itself. However – and this is unfortunately overlooked by Latour – time plays a crucial role here. There is always a time-lag between ordering and revealing. Resistance is a slowing down, an inertia. The speed of ordering is kept in check by what is being ordered. For Latour, this time-lag is more or less constant. For Beck it is increasing. Yet, at the same time, through anticipation, which is itself historically produced by the acceleration of feedback loops (learning and memory), the present is extending into the future (Nowotny, 1994). Risk is itself an anticipation (that something bad might be happening), thus bringing the future into the present.

The risk society thesis suggests that the inertia of resistance becomes more problematic when the force of immutable mobiles is accelerating. That is to say, the gap between inert reality and accelerating trials increases to a point where anticipations of risks start to coincide with their manifestations as plausible accidents, disasters and catastrophes. This presents itself in a sentiment that technoscience is increasingly losing control over reality. We now anticipate that most technoscientific innovations and interventions will have unintended consequences. By mere means of our anticipation these unintended consequences have already become risks before they are hazards.

Moreover, this notion of technology brings a second and quite specific intervention to understanding objectivity. By representing, one is always reminded of something else, of the absence from which the presenting is drawn in the process of objectification. This relates to Derrida's (1982) 'concept' of *differance* as a critique of 'the metaphysics of presence'. The metaphysics of presence suggests that Being and Presence are the same '(no)thing', as if materiality coincided with an unmediated and immediate 'present/presence' (Van Loon, 1996a). Differance points towards that which differs and is being deferred; it points towards a necessary remainder of that which cannot be brought into full presence, of the traces of the fabrication of the present, of the forgetting of the difference between being and Being. However, metaphysics of presence is – in a way – an inevitable part of differance, that it is this differance that seduces us to forget the difference between sign and referent and between being and language.

It is possible to connect Derrida's writings on differance with Latour's notion of reality as resistance to trials. If differance refers to a displacement, an impossibility of full presence, then it is similar to the resistance through which 'reality' manifest itself during trials. The technology that orders and reveals a virtual object as real thus realizes this objectivity through various trials. These trials take place both in laboratories and in 'the field', and are further extended to the domains of rhetoric and discourse as means of 'fixing' specific facts as no longer mere constructions. The constitution of specific networks is based on linking specific trials.

However, this does not necessarily require all the trials to 'add up'. Indeed, the times when different trials actually do add up (as in the ideal typical 'normal science' described by Thomas Kuhn (1970) in terms of paradigm) are quite rare. More often than not, different trials invoke different kinds of resistance. That is, they enframe and reveal different kinds of 'virtual objects'.

John Law (1995) uses the example of 'accountability', which is enframed and revealed in an heterogeneity of performances throughout a particular organization. These performances are sometimes enacted by the same and sometimes by different actors. For example, in the management of accounting systems, accountability relates to an empirical understanding of 'the account' as corresponding to an actual financial reality 'out there'. However, accountability becomes something different in bureaucratic procedures where it is enframed as a tool or instrument of processing energy and particle flows most often paper work (Law, 1995: 284). Within these two different practices, accountability may be further split into a range of virtual objects, for example systems of tracking and monitoring, procedures of enhancing 'transparency', increasing calculability, evaluability and of course enabling judgement. Moreover, Law argues that these different practices engage different kinds of subjectivity and different senses of action and intervention.

In a similar vein, Annemarie Mol (1998) gives the example of the diagnosis and treatment of *atherosclerosis*. For her, the disease is primarily enframed/ revealed in medical work (even if it relies on self-diagnosis of the patient). For her a successful diagnosis and treatment is about making effective links between complaints and what medical textbooks describe as the clinical definition of atherosclerosis: 'the thickening of the intima of the vessel wall' (Mol, 1998: 146). She argues that on a basic level there are two ways of diagnosing atherosclerosis. The first is clinical, it involves a medically trained professional, usually a doctor, asking questions and the patient answering them, often followed by an external examination of the affected body parts. The second is an assessment of a sample of the blood vessel wall, outside the living body, in a pathology laboratory and generally done through microscopes. The links between these two trials are performed through imagination and deliberation; that is, the ordered/revealed 'virtual object' becomes 'fixed' in specific semiotic settings. However, the link between the two virtual objects is finalized only if the body part is removed from the human being, that is through amputation. It is obvious that even in cases where this is possible, doctors would be extremely cautious in taking such drastic measures for the sake of a semiotic closure of a virtual object, unless the patient is in severe pain and no alternative treatment has been effective. Hence, clinicians deploy a range of additional technologies to reveal and order the atherosclerosis virtual object such as stethoscopes and Doppler tests, which 'index' the severity of the narrowing down of the blood vessels.

In these examples, we see how as a general feature of ordinary everyday practices, the enframed/revealed 'virtual object' remains unstable and multiple. Moreover, they show the importance of intersections between technologies, practices, subjectivities and responses (to trials). None of these enacts reality-effects on its own. The multiplicity of the virtual object, however, is not generally conceived of as a problem in itself, although problems of linking sometimes do emerge. Somehow, most actors seem quite competent in negotiating between different virtual objects in their orderings of everyday life. Hence the 'reality status' of virtual objects may be unstable and multiplied across practices, this does not significantly affect the day-to-day business of technoscience.

Risk as a virtual object: the case of BSE/vCJD

It is at this critical edge that 'risks' could be theorized. Although Latour and his associates rarely discuss risks as such, I have argued in the previous section how their understanding of technoscience is quite adept to understanding risk. If we take for example 'risk-assessment practices', we could follow Law and Mol to suggest that 'the risk' itself is constituted as a virtual object. Take BSE/vCJD for example, we could see a multiplicity of risks emerging in different risk-assessment practices, in which the collusion between government, the agricultural industry and a significant section of scientific institutions could be understood as a process through which one particular understanding of risk of humans contracting vCJD from consuming (BSE-contaminated) beef was for a long time fixed, even if confronted by various trials which enframed and revealed a different kind of risk.

First of all, in technoscience, BSE and vCJD are both classified as 'transmissible spongiform encephalopathies' (TSEs), which share with other types of BSE that they transform the cellular structure of brain cells into a sort of sponge, through mutations of protein cells. The disease pattern of classic CJD – which is found around the world – consists of a progressive degeneration of the body through the rapid and irreversible destruction of brain tissue, causing death. It is part of a wider group of human and animal diseases which are all marked by distinct patterns of electrical activity in the brain (Beauvais and Billette de Villemeur, 1996). Although brain scans (EEGs) are able to reveal such patterns and point towards this disease, the definite presence of the disease can be established only after autopsy, once the victim has died. The brain is left riddled with holes like a sponge.

According to the UK government's Department of Health, variant CJD is different:

> This variant has not been previously recognised and is characterised by behavioural change, ataxia, progressive cognitive impairment and a tendency to a prolonged duration of illness (up to 23 months). The EEG is not typical of classic sporadic CJD. Brain pathology shows marked spongiform change and extensive florid plaques throughout the brain. As with typical sporadic CJD, symptoms in the early stages may be fairly non-specific.
>
> (http://www.open.gov.uk/doh/cjd/cjd_pubs.htm)

However, for public health concerns, more interesting than ascertaining whether someone has a TSE or not, is the question what causes it. As the actual pathogen of the disease is still a mystery (Ford, 1996), there remains a wide range of conflicting hypotheses.

In such a situation 'normal science' responds with indecision. It cannot act. This is the job of governments. The UK government did act, albeit hesitantly, with large intervals and on the basis of a rather selective reading of information provided by technoscience. That is to say, the immutable mobiles that operated between technoscience and governance were rather limited in the kinds of virtual objects they en-presented. Before a link between BSE and vCJD was no longer denied (in March 1996), the Conservative government in the UK stubbornly held the view that BSE was only a disease for cows (Phillips, 2000). It thus took more than a decade after the first serious scientific warnings about the risks of eating BSE-infected beef for public health, before the government decided to take a slightly more conservative approach to the possible risks, by allowing a different hypothesis into the frame, namely that a link between BSE and vCJD could be possible (Toolis, 2001).

Was this neglect down to pure incompetence? Perhaps, but the government's virtual object of BSE/vCJD may have been rather different from that of technoscience. The government's risk assessment did not solely concern the world of brain cells, but was extended to include farming, the beef industry, the food industry in general, and of course, risks related to the more limited internal worlds of politics and governance (Adam, 1998). The collapse of the beef industry and cattle farming may have been perceived as greater risks than the hypothetical infections of a few unfortunate human beings (Macnaghten and Urry, 1998).

However, after their humiliating climbdown of 1996, the Conservative government had to face more threatening risks. Even before this event, the media were already haunting politicians with reminders of more negative scenarios offered by scientists other than those allowed to represent 'Science' by the Department of Agriculture. Now politicians were confronted with additional political risks related to their neglect of public health concerns. The media themselves too faced some risks, as they had to balance market pressures with their proprietor's editorial policies. Thankfully for both government and media, a new scapegoat was quickly found. The European Union (EU) entered the scene by enforcing a ban on the export of British beef worldwide. By acting where the UK government failed to act, the EU also exposed the real decline of national sovereignty, especially in the economic and political spheres of risk management. It thereby exposed new risks of a decreased autonomy of the British state, against which populist media and politicians were able to redirect some of their anger and concern. The enrolment of the EU thus transformed some risks into opportunities. Just as the media could cash in on the scandal and sensationalism of political failure, so could both media and politicians cash in on populist nationalist sentiments *vis-à-vis* 'foreign interference'. Amidst all this turmoil, in which risks and opportunities multiplied, retailers and producers were also enrolled in risk-taking strategies. Supermarkets launched campaigns in defence of British beef, whereas local (organic) butchers for example could claim that their beef came from BSE-free cattle.

What is important to note here is the way in which the BSE/vCJD risk transformed and multiplied into a range of virtual objects – political, industrial, economic, administrative, representational risks emerged alongside the epidemic ones. Eating beef became more politicized than ever before. Another key conclusion is that these risks do not add up, and that there were no immutable mobiles that functioned to integrate all actants into a single network. The multiple worlds of BSE/vCJD remain divided until this very day. As yet, we have not seen a phenomenon similar to that of 'Pasteur' in late nineteenth-century France that could combine many of these disparate elements into a single model. The multiplication of unstable virtual objects could thus be understood as a risk of destabilizing actor networks in a breakdown of intersystemic communication (translation), without readily available immutable mobiles to help repair the damage.

Conclusion

ANT enables us to see how different elements connect. It also allows us to understand the work of technoscience alongside other social systems. The concept of immutable mobiles furthermore helps us to understand how specific translations are made, and turned into 'representational models'. Finally, we have seen how risks can be understood as 'virtual objects', whose presence remains rather inconclusive, and fully dependent on the technologies of ordering/revealing. The great advantage of conceptualizing risk as a virtual object is that it highlights that the process of risk formation involves a number of crucial stages of translation, which coincide with the inauguration of specific actor networks based on risks (also see Chapter 5).

First, virtual objects such as risks have to be visualized, which is more than the provision of images; they also require an imaginative actualization. Second, these virtual objects have to be named, that is, signified, to develop a 'parallel' practice of discursive mobilization. Third, they have to be valorized, that is, their 'meaning' has to engage particular exchange relationships (or cycles of credit as Latour and Woolgar, 1979, called it), be it economic, political, symbolic or moral. This shows that formation of risk as a virtual object is a deeply political process, which also has ethical implications which extend beyond the domains of technoscience.

However, there are some noticeable shortcomings in ANT and especially Latour's work. First, whereas Latour's work has been very effective in exposing closures, stabilizations of networks and fixations, it is much less adept to work on the second kind of technological culture, i.e. that of systemic breakdown. Latour's anti-historicism backfires in his inability to acknowledge 'time' as a constitutive factor of representation. The essence of risk, however, is time, as it marks a future projected onto the (extended) presence through anticipation. Because he fails to consider both inertia and acceleration as volatile and historically specific, Latour's own version of ANT does not support a theory of dysfunctionality as well as, for example, Law and Mol.

Second, and related to the first, whereas Latour's work provides an excellent perspective on how to understand 'the bringing into presence' of risks, it does, however, provide fewer clues as to how to understand the 'dark side' of that same process – its concealment. ANT is predominantly concerned with 'presence' in the form of actants, instruments, practices, objects, etc. It thereby has less ability to consider silence or absence as anything beyond not-present. For example, the concept of 'virtual object' could be seen as an attempt to theorize absence, but there always remains the assumption that there is some 'thing' that is being virtualized and objectified. More specifically, when addressing the crucial issue of motivation, ANT is forced to limit its understanding of political sensibility to forms of consciousness and intentions. Because the logic of science is inherently bound up with the construction of 'matters of fact' (as black boxes), it is simply not attuned to thinking beyond the actual, especially with reference to its own works of mystification. Concealment is beyond the remit of ANT because it is entrapped by a metaphysics of presence. However, its dislike of 'differance' makes it more difficult to render an account of absence beyond that of 'virtual presence'.

A third weakness in Latour's work is that he fails to adequately discuss the issue of motivation. Whereas the socio-logic performs the bottom-line argument for Latour, this logic appears rather 'thin' when we consider the motivational aspects of 'enrolment'. It seems that at best, the motivation is the enhancement of life chances, which might manifest themselves as either self-interests or collective interests (or both). That is, the connection of multiple 'fates' under a particular mode of rationality of technoscience is driven by both an anxiety over death and a lust for life. That this latent Darwinism makes him a close ally of structural functionalism, and perhaps above all Luhmann, is however rather at odds with other aspects of ANT.

Of course, as he himself has been trained as an anthropologist, Latour would be the last to neglect the cultural embedding of such anxieties and desires. However, as an ethnographer who is interested in science-in-action, in forms of the actual practicing of reason rather than tracing its materialization and in the regularities of action patterns, he does not conceptualize that which is not present-at-hand in practical reason. In other words, he falls prey to the same metaphysics of presence (conflating the actual with the present) that he seeks to avoid. For example, his opposition to 'the unconscious' or any related concept that attempts to incorporate 'absence' (which is simply reduced to 'action at a distance'), makes it rather difficult to think of modes of rationality that are not primarily invested with 'deliberate strategies'. Enrolment seems to require some form of mediation in terms of awareness. Consequently, it is insufficiently clear how such cultural embedding of anxiety and desire operates for the non-human actors in the network; unless they are being anthropomorphized into being similar consciousness-bearing creatures. What are the strategies employed by technologies, non-human organisms and gods? And why do they deploy these? These questions will be the subject of the next chapter.

4 Assemblages and deviations

Biophilosophical reflections on risk

Until now, we have discussed risks as fairly unproblematic. Whereas we can only speculate on the reality of risk beyond their articulation in (resistance to) trials, it is hard to deny that risks exist, if only because actants actualize them in everyday settings. Such an assertion of the existence of risks thus follows the now rather uncontroversial (weak) 'social constructionist' route (Lupton, 1999) found, for example, in the works of Douglas and Wildavsky (1982), Luhmann (1982, 1995) and to some extent Beck (1992, 1994, 1995, 1997, 2000) and Giddens (1984, 1990, 1991, 1994). Weak social constructionism is based on an agnostic stance towards the relationship between being and perception. That is to say, it defers judgements on the question whether something is real or not, by focusing instead on how people make it real (actualization). It differs from realism, for which there is a real outside of our perceptions of it, which manifests itself through our worldly experiences and which to some extent governs how we can experience reality (even if only through distortions). It also differs from stronger versions of social constructionism for which there is no reality outside perception (sometimes this is equated with discourse or 'text'). Weak social constructionism simply suspends the difference between perception and actualization. In other words, it makes no sense to distinguish between the virtual and the real, because risks are always virtual. By their very nature, risks cannot be 'real', for they would cease to be risks if they were more than a potential of becoming (Van Loon, 2000a; Beck, 2000).

In the previous chapter Actor Network Theory was used to conceptualize risks as virtual objects. As virtual objects, risks are both fragile and multiple. This chapter will further analyse this process of objectification by highlighting its biopolitical nature. That is, the fragility and multiplicity of risk is dispersed across and disseminated by actor networks as assemblages of power and desire. Rather than suggesting that they are merely contingent outcomes of arbitrary forces, set into motion by actants who are (indifferently) exercising a will to power, a further exploration of the issue of *motivation* is necessary. The technological assemblages involved in ordering and revealing risks also transform the arbitrariness of risks into a more concerted and organized abstract machine of governance and control. This is the price we pay for our liberal-individualist hegemony. Risk becomes opportunity, anxiety becomes desire and scarcity becomes excess.

However, whereas it would seem almost self-evident to pinpoint to (postcolonial) capitalism or patriarchy to identify the continuation of interest-specific valorizations in the risk society, we should also acknowledge some lessons from the previous chapters. The first comes from the risk society thesis and shows us that risks and opportunities are not distributed in the same ways as scarce resources. The second comes from Actor Network Theory and stresses the need to look at multiple strategies of network-building and incorporation, which may link different kinds of interests in rather unexpected ways.

This is to say that science, governance, mediation, commerce law and military institutions may not always — and in fact rarely do — add up smoothly to become one system in service of particular class, ethnic or gender interests. Examples such as the gay activists in the field of HIV/AIDS, environmentalists in the field of conservation and feminist movements in the field of family planning show that under particular circumstances, networks can emerge in which seemingly powerless and marginal groups have successfully seized quite a nodal position in specific political domains. This often happens through specific associations between technoscientific and legal expertise, political strategies, public relations and entrepreneurial initiatives that have taken elements from the mainstream and reorganized them in specific emergent assemblages that have territorialized previously unchartered domains.

An analysis of social groups acting on the basis of interests must place central emphasis on intentions and consciousness. It depicts a world of collectivities who know who they are (who belong to 'us' and 'them'), what they want and how they want to get it. They must conceive the world in terms of a struggle in which 'their' interests are opposed to those of others. Hence, the emphasis is on rational, mental and cognitive processes for which emergent risks and opportunities are merely incidental contingencies which may either help or hamper the realization of their objectives. Even in various forms of Marxism, which do allow for some sense of latent interests, class struggle always involves a transformation of latent interests and needs into manifest ones. Without that transformation, the status quo will remain — because the powerful do not have latent needs and interests; they always-already know them.[1]

As soon as we take interests and needs of collectives as central; as soon as we allow a picture of the world carved up in these neatly categorized and labelled antagonisms, we lose the sense of 'reality' as that which resists trials. Instead, our labels and categories take primary stage and we are forced to think through them in representational terms. That is, we have to accept the meaning of representation as equal to that of re-presentation; making present becomes identical to replacing. In the previous chapter, we have seen how in emphasizing the difference between re-presentation and representation, Latour (1998, 1999a, 1999b) separates scientific truth claims from those made by religion. He thereby acknowledges that the trials that are being generated by science and religion are irreducible to each other, and hence the realities that resisted those trials remain incompatible. That is why scientific criticisms of religious truths have little effect on the religious faithful and vice versa.

In this chapter, we will explore in greater detail the theoretical grounding of Latour's critique of reducing re-presentation to representation, in terms that are not openly acknowledged as part of ANT, but which can be argued to share very similar philosophical origins in phenomenology and hermeneutics. This has important implications for the way in which we theorize risks as virtual objects, and more specifically, for the kind of biopolitics that this engenders.

The issue of motivation takes centre stage here. If we take seriously the radical implications of the risk society, namely that risks are inherently ambivalent in terms of their implications for the ordering of society, we cannot explain the logic of social ordering by pointing towards whoever seems to be benefiting from it. Capitalism was not created and sustained by the bourgeoisie because they knew who they were, what capitalism was, or how it could be created and sustained. It emerged from specific practices that made sense under those specific circumstances at those specific times. There was no grand architect; it was an accident bound to happen, destined to happen perhaps, but not because any single individual or group wanted it to happen before it happened. Even today, one could argue that it is not the bourgeoisie who sustain capitalism, they are not able to do that. Capitalism is simply too complex. As Gramsci (1971) wrote more than 70 years ago, the bourgeoisie have always been unable to take care of the long-term interests of capitalism because they cannot act in concert (an effect of the competition imperative), they cannot plan to act for the long term because their timescales do not exceed months, let alone years.

At first sight, however, patriarchy may be a bit more easily seen as an off-spring of man's consciously articulated sexist desire for domination over women, as are systems of colonialism based on racist, ethnocentric and xenophobic ideologies. Indeed, unlike capitalism, these systems do not seem complex at all. However, what is more difficult to explain is how they have historically evolved, and why – despite nearly a full century of liberal pluralist domination, women and people-of-colour still find themselves in systemically sustained marginal social positions. Again we will have to look to factors beyond ideology and intentional motivations, and seek explanations in terms of the complexity of social relations which are governed by unconscious as well as conscious desires and forces operating beyond the immediate will to power articulated by individuals and groups.

The point here is that rather than focusing on 'interests' which are being served in more or less conscious manipulations of institutions and networks by organized individuals, we should continue our quest using the Latourian notion of 'forces', which do not require articulation in intentions and awareness, but may have more autonomous motivations. This is not to say that there are no interests or needs, or that the way in which networks are arranged does not benefit some groups in society more than others, but simply that the arrangements into which actants enter particular social orderings are not constructed according to strategies that articulate those interests. Of course there are interests, of course people act on them, but these actions are assembled into complex networks and flows, where interests are connected to other types of forces such as unconscious desires and unreflective motivations. Interests constitute but one vector in a highly dense and multiple assemblage of flows, and perhaps not the most revealing one.

Rather than interest-based politics, we need to focus our attention to biopolitics to understand more fully the implications of a non-humanist (but not anti-humanist), technologically indeterminate conception of risk. In order to move beyond the cognitivist bias of motivation, we shall shift the focus to what has been called 'biophilosophy' and more particularly, some ideas offered by Haraway, Deleuze and Guattari and Lyotard. It will be argued that biophilosophy enables us to conceptualize the technocultural ordering of risk as a 'normative' issue.

Politics, embodiment and technoculture: diffractions of Donna Haraway

One weakness that is often associated with ANT is its lack of concern for the political moment of technological culture. That is, whereas it has proven to be a very useful framework for describing the establishment of truths and facts, it is less helpful when trying to understand the political implications of such technoscientific enframing. However, it is wrong to suggest that ANT has no sense of politics. Indeed, for Latour, as well as for others associated with it, such as Callon, Law and Mol, actor networks are political accomplishments. The problem with Latour's sense of politics, however, is that it is often reduced to Machiavellianism and principally concerned with 'effects' rather than 'principles' (it is procedural rather than substantive). That is to say, because there is no scope for the unconscious, the articulation of forces in social settings becomes purely a matter of strategy; structural interests may have disappeared, they have been simply replaced with more local and strategic ones. Hence, scientists operate as scientists because they want to achieve international acclaim, fame, recognition by peers, perhaps a Nobel Prize, in any case, accumulate credit (Latour and Woolgar, 1979).

Of course, this is more accurate than stating that scientists simply want to serve capitalism, or even that they want the highest possible salary. However, because ANT prohibits any generalizations on the basis of pre-established sociological labels, such as that of the 'science' profession, one rarely comes across any statement of interests, motivations or intentions of any of the actants, except that they want to maximize strength (e.g. credit, stability and influence) through the setting up of networks. There is no other kind of force at work than those inducing or depleting strength of actants. There is no sense of beyond, of infinity, or of morality. Ethics is reduced to simply rules of engagement and attachment. Hence, there is little evidence in ANT that actants could withstand cynicism or bad intent if this would deliver increased strength. In the face of risks, and its correlate in anticipated suffering, the question of normativity in technological practices, however, cannot be avoided.

It is the work of Donna Haraway that can be seen as one of the most sustained and explicit attempts to cast the enframing established by technological culture within the problematic of 'normativity' on a macro-political scale. This normativity can be revealed by the originally Leninist question 'what needs to be done?', but is now freed from the programmatic aspects of economic determinism. That is to say, for Lenin, it was always obvious by whom politics needed to be done: the leaders of the proletarian revolutionary class. For Haraway, however, the 'by whom' becomes a most significant addendum. In her writings, a number of actants have featured: cyborg, coyote, material semiotic-actor, and more recently, *Modest Witness* (Haraway, 1997). A key feature of all these actants is that their politics is always complicit with the world they intervene in. Although the actants may be marginal, they are never totally 'outside'.

> My modest witness cannot ever be simply oppositional. Rather s/he is suspicious, implicated, knowing, ignorant, worried and hopeful. Inside the net of stories, agencies, and instruments that constitute technoscience, s/he is committed to learning how to avoid both the narratives and the realities of the Net that threaten her world at the end of the Second Christian Millennium. S/he is seeking to learn and practice the mixed literacies and differential consciousness that are more faithful to the way the world, including the world of technoscience, actually works.
>
> (Haraway, 1997: 3)

When we bring the issue of the agency of normativity in contact with the notion of technology as ordering and revealing, the question arises of how the agency of technological normativity is enframed, or perhaps better, incorporated.

Incorporation must be seen in two ways. First, there is incorporation in the sense of 'becoming part of'. One could be incorporated into a society (become a member), form of government or religion. Most important for the modern world is capitalist in-corporation, which refers to the logic of commodification. Capitalism absorbs all other forms, be it human beings, objects, economic systems, forms of exchange, or ideologies, and turns them into expressions of capital – it imposes the commodity form onto all recognized entities.[2] This imposition is the first requirement of entering an exchange relationship or capitalist enterprise. Incorporation under capitalism is thus a specific form of 'valorization'. This type of incorporation thus works on the basis of an usurpation of the core form of value (capital) which provides the code through which everything else becomes 'transformed' as an expression of that particular value system through 'exchange relations'.

However, technology entails a second form of incorporation, which refers not to the becoming part of, but to the part-becoming, or 'embodiment'. In this sense, incorporation refers to a process not of usurpation but of 'induction'.

Transformation is then a becoming-flesh (in-carnation) or becoming-material. Of course, this process is related to the first. When becoming a member, one not only takes part, but also becomes-part; one incorporates the ethos, the habitus of the collective; one forms an organizational body – a corporate-corpus. In early-modern capitalist society, this incorporation took place in the form of discipline. The becoming-corporate was a matter of training and supervising, regulating and monitoring, the formation of particular postures and attitudes (Foucault, 1977b, 1979, 1980). Discipline then, could be seen as the counterpart of valorization. Together, discipline and valorization constitute the material-semiotic organization of modern being.

Technology, incorporation and embodiment form the backbone of Haraway's biophilosophy. However, cutting across the aforementioned theoretical spectrum is the concept of gender. Haraway's biophilosophy is gendered, and it is gendered difference that constitutes her understanding of technological normativity. For Haraway, (human) incorporation is essentially gendered. Hence valorization and discipline are equally framed in terms of engendered difference. This uncannily echoes the implicitly gendered universalist assumptions of modern technological culture as a triple exercise of domination of 'man' over nature, divinity and history.

In *Primate Visions*, Haraway (1989) fundamentally questions the distinction between nature and culture, by looking at what seemingly lies in between: the world of primates. The power of primatology as a discipline for understanding human nature lies in its capacity to project and enforce particular stories and accounts of human being onto its object of study: primates. The writing of these stories, often aided by particular technologies in fieldwork and laboratories, produces a whole apparatus, discursive formation and set of practices, to order and classify the world onto a singular grid which is gendered and racialized. Not surprisingly, the White Caucasian man stands at the centre of that grid; all differences are deviations from that pure model.

For example, it is quite clear how the gendered divisions of power are structured when looking at reproductive technologies. The application of medical knowledge and appliances is generally performed by men onto the bodies of women, to exercise a particular control over the reproductive process. Just simply looking at the medicalization of childbirth makes one acutely aware of the shift of control over the birth process away from the mother towards an institutionalized machinery where men are usually in control (Franklin, 1997; Martin, 1987). Questions concerning scientific knowledge should therefore always take account of the hidden politics, which are both gendered and racialized. However, these politics are not necessarily following deliberate constructions and intentional motivations. Haraway never loses sight of the technoscientific ordering which follows a logic that is relatively autonomous from the intentions of those that engage in it. She does not question that primatologists actually see apes behaving within heterosexist patriarchal modes of ordering. And she does not assume that they see this only because they want to, or because they deliberately erase alternative explanations.

One particular concern raised by Haraway in this respect is that of 'vision'. Science is often metaphorically modelled on the notion of the distanced, disinterested gaze.[3] Science is based on 'observation' – what she calls 'a persistence of vision'. The problem with this modern aesthetic is that a disinterested, disengaged 'vision' is also disembodied – it is attached to no body in particular. This disembodied notion of scientific practice all to easily slips into a forgetting of the desires of the body, which are sexualized and gendered. A model of science based on disembodied vision ignores that vision is a highly sexually charged activity, loaded with investments of particular (fetishized) desires (Jordanova, 1989).

Haraway (1988) extends this critique to an understanding of 'knowledge' in general. Instead of rejecting science as a technology of ordering, she argues that feminism should reclaim some of its ground by asserting possibilities for objectivity not based on universal disembodied vision, but on located, embodied, particular perspectives (technological incorporations) against the valorization of the mainstream disciplines. However, rather than claiming a superior ground for a general 'standpoint' based on the universality of women's oppression under patriarchy (e.g. Smith, 1974), Haraway advocates a 'partial perspective'. 'Only partial perspective promises objective desire.'

However, she goes further. In the world of today, vision has been technologically extended by various media, which have produced the paradoxical consequence that at once, it has allowed a further disembodying of the gaze (surveillance) but also intensified the visibility of desire as ubiquitous in (post)-modern western cultures. In other words, the omnipresence of technologies of visualization and representation has created a momentum in which the disembodied gaze of the White male scientist has lost most of its credibility (also see Denzin, 1995). The body is now completely interconnected with technologies and cannot claim to exist independently any more. Indeed, as the body dissolves in technological systems, 'vision' also loses its centred perspectivity. Through, for example, digital imagery, the assumption of equivalence between 'optics' and 'iconicity' has become untenable (Lury, 1998). Even for the most uncompromising forms of technoscience, seeing is no longer believing.

Haraway takes the idea of a connection between body and technology to its logical extreme in the *Manifesto for Cyborgs*, which was originally published in 1985. Whether we like it or not, she argues, we have become cyborgs. Through the process of technological incorporation, the distinction between human being and machine has been shifted and blurred radically. The cyborg heralds the end of all originary purity, all claims to an indivisible 'origin' – we are all hybrid beings. Cyborgs live in a world of simulacra: neither reality nor representation provide adequate models of engagement with our environments, because both still invest ideals of a purity of origin, and the possibility of a final unity between words and things. The cyborg is thus a model for embodied vision and situated knowledge. With the cyborg as a model of survival, it is clear that our future existence is no longer conceived as one of human purity. However, although from a humanist perspective this might sound rather apocalyptic, the apocalypse itself heralds a promise of another future. Cyberfeminism is one particular way of attempting to rewrite the present to reclaim the future.

Haraway's widely acclaimed cyborg-feminism is now more than 15 years old; it is still remarkably adequate in theorizing the politics of technocultural complexity as it transgresses the boundaries between the organic and the technological – the body and the gadget. Although Haraway's approach does not fit easily into any category inside or outside the various feminisms, her criticism of enlightenment rationality, including humanism, and her equally critical but playful stance *vis-à-vis* structural fixations, have aligned her work more closely to some approaches associated with (feminist) postmodernism and poststructuralism (Van Loon, 1996b). However, what makes her work different from pure postmodernism is her continued defiance of nihilism, often expressed in terms of hope, or a desire for 'progress'. Moreover, she cannot be subsumed under poststructuralism because although she recognizes the failures of subject-centred identity-thinking, she does not let go of a need for a politics of (non-isomorphic) subjectivity. That is to say, her continued involvement in a socialist-feminist and postcolonial agenda means that she cannot avoid referring to systems of power that take on structural forms: capitalist, patriarchy and western-global imperialism.

However, this does not mean that she is willing to reduce her analyses of technoscientific practices to such macro-systemic forms of organizing interests and subjectivities. Instead, she investigates ways in which the military apparatus, hypercapitalist markets, the global ecosystem and multimediated science culture have produced illegitimate offspring who have dispersed and displaced some of the essential 'genetic' information upon which these apparatuses are built (Haraway, 1997: 12–13). As in her earliest writings on the subject she uses the notion of cyborg as a figure into which a whole universe of (embodied) being in the world has imploded, in a similar way as the atom is an implosion of *physis*, the gene is an implosion of life, and the bit is an implosion of information.

What the implosion signifies is at once a mode or re-presentation or revealing, and a mode of representation as ordering, while maintaining a difference between the two (see Chapter 3). What this means is that whereas the figure of cyborg, gene, atom or bit can be understood as metaphors and even allegories in a semiotic configuration of life at the end of the Second Millennium, after the dreams and promises of industrial modernity have totally evaporated, they are also 'performative images that can be inhabited' (Haraway, 1997: 11). They are simultaneously literal (performative-material) and figurative (semiotic). The material-semiotic actor or modest witness is thus always-already split; there always remains something in excess of what is either enframed or revealed which makes full closure forever deferred to an infinite horizon. This is also why the witness must be modest; any claim to totality shall be in vain.

The doubling of matter and meaning enabled modern scientists to create experiments and later to engage in developing scientific logic, while assuming independence from the ruling political or religious doxa. The scientists bore witness to nature unveiling itself through controlled experiments. Whatever was being observed were 'matters of fact' about which neither state nor church could do anything. However, whereas modern thought sought to reassemble the outcomes of experimentation with a more encompassing and universal logic of objective reason, to close – once more – the gap between what is revealed and what is enframed; it has never succeeded in doing so (Latour, 1993). From the moment of separation of reason from nature, there was no way back. Either one had to accept that reason is unnatural, as for example Kant had to concede towards his death, or that nature is unreasonable, as for example those trying to prove Darwin's (or Einstein's) postulates have painfully discovered.

It is this deviation from the closure of enframing and revealing that stands at the core of Haraway's conception of the politics of technoscience. This deviation could also be used to disseminate the notion of 'risk' across a more multiple and diffused domain than that of the institutions of modern society. That is to say, many reflections on the concept of risk have returned to the problematic articulation of reality and perception, leading to a wide spectrum of realist and constructionist 'risk theories', who each give different emphases on either side of the relationship (see Lupton, 1999). What most of them have failed to do, however, is to explore the connections between 'reality' and 'perception', and more specifically, how these connections are produced or induced.

The need to overcome this painfully unhelpful dualism has always informed Haraway's work. Long before *Modest Witness*, she already reflected on a particular figure to understand the materiality of representation and vice versa: the immune system.

> The immune system is an elaborate icon for principal systems of symbolic and material 'difference' in late capitalism. Pre-eminently a twentieth-century object, the immune system is a map drawn to guide recognition and misrecognition of self and other in the dialectics of Western biopolitics. That is, the immune system is a plan for meaningful action to construct and maintain the boundaries for what may count as self and other.
>
> (Haraway, 1991: 204)

The immune system deals with matter out of place: cells, proteins, molecules that do not belong, that are marked 'other' and deemed threatening. In her essay on the biopolitics of immune systems, Donna Haraway (1991) discusses the way in which concepts of immunity have emerged within scientific discourses and have spread to wider discursive formations that constitute the social. Triggered by the relatively recent discussions around AIDS as an epidemic of metaphor (Treichler, 1988), Haraway restates her well-developed argument of *Primate Visions* (1989), that science is an elaborate technology of representation which finds its roots in the everyday conceptualizations of the order of the world; but it is also a representational practice that helps to reconstitute the very common sense that such conceptualizations make.

Whilst as such this claim is quite commonplace and can be found in quite a range of approaches in the sociology of scientific knowledge and science and technology studies, Haraway's position is more radical than this. Not resting her case with the (constructionist) assertion that science is a representational practice, she inserts a rather different argument. Instead of focusing on correspondence and representation, she is concerned with the materialization of representational practices.

> The immune system is both an iconic mythic object in high-technology culture and a subject of research and clinical practice of the first importance. Myth, laboratory and clinic are intimately interwoven.
>
> (Haraway, 1991: 205)

Hence, at the core of the problematic of the biopolitics of post-modern embodiment is not the question of resemblance or equivalence between scientific and symbolic representations, but the very interconnectedness between scientific and representational practices. This connectivity is what Deleuze and Guattari (1988) referred to as the *assemblage*:

> [T]he principle behind all technology is to demonstrate that a technical element remains abstract, entirely undetermined, as long as one does not relate it to an *assemblage* it presupposes. It is the machine that is primary in relation to the technical element: not the technical machine, itself a collection of elements, but the social or collective machine, the machinic assemblage that determines what is a technical element at a given moment, what is its usage, extension, comprehension, etc.
>
> (Deleuze and Guattari, 1988: 397–8)

In the previous chapter, the concept of virtual object was used to refer to an almost identical argument. The assemblage is a shifting and multiple technological constellation which enframes and reveals virtual objects through what could be called 'ordering' or 'challenging forth'. Ordering as both putting something in its proper place and commanding thus refers to the active process of selection and articulation (and de-selection/disarticulation). Hence it is not the virtual object, the technical element, or any particular technology of objectification (e.g. the electron microscope), but the entire constellation of virtual objectification that constitutes the machinic assemblage of technological culture.

We need to keep this at the focus of our attention when subsequently encountering Haraway's argument that the immune system is based on a presupposition of identity. That is, in order to form antibodies, the body's immune system must be able to distinguish between what belongs and what does not. Like the diagnostician, who must be able to identify a pathogen as virtual object, for example as accountability, atherosclerosis or vCJD, the immune system itself deploys a mapping of the molecular structure of the body to identify possible risks to its integrity. Hence, the immune system must be able to 'identify' difference and thus implies representation. It is an apparatus that creates and manages boundaries.

However, what counts is not the symbolic manifestation of the identifications that immune-system-discourse disseminates, but the very practical materialization it generates when becoming input for information processing systems. Haraway's notion of 'the body' as a material semiotic actor operates at the core of this understanding.

> Bodies as objects of knowledge are material-semiotic generative nodes. Their boundaries materialize in social interaction; 'objects' like bodies do not pre-exist as such.
>
> (Haraway, 1991: 208)

Hence, far from a mode of representation, the immune system is a technology that assembles and inaugurates 'entities' within the classification system of 'self' and 'other', which, because it is still linked to a representational practice, also becomes a question of 'same' and 'different'. In other words,

> while the late twentieth-century immune system is a construct of an elaborate apparatus of bodily production, neither the immune system nor any other of bio-medicine's world-changing bodies – like a virus – is a ghostly fantasy.
>
> (Haraway, 1991: 209)

The matter of embodiment is important to Haraway, because it signifies the limits of bio(techno)politics. Although these limits are not confined to those of the human organism, any node of application will both inhabit and be inhabited by material semiotic actors. The immune system may be technologically enhanced (through cybernetic feedback loops for example), and may be disseminated across a network of information flows, it never ceases to differentiate between what belongs and what does not, between attachment and detachment, between subjection and abjection, and this differentiation is never a purely instrumental operation but a matter of greater significance. That is to say, the immune system deploys technologies of normalization and entails a sense of normativity.

Whereas ANT is mainly concerned with valorization as a particular *effect* of rationality, which serves a merely strategic purpose of enhancing/maintaining strength, Haraway extends the notion of 'value' to a more cultivated and embodied sensibility that can be traced in the flows of ethico-political movements, such as feminism and eco-socialism. Hence, technological culture is not devoid of questions of justice and responsibility, although the forms which these issues take are far less determined than they used to be. That is to say, taking the example of 'immunity' politics, gender may cease to be a significant genre of differentiation, but the organization of the reproduction or engendering of the species will not be; race may cease to be classified in terms of stock or skin colour, but genetic differentiation will still be significant for military, financial, legal, medical and cultural systems of ordering and boundary regulation.

This echoes some ideas offered by Douglas and Wildavsky (1982), for whom risks were mere devices for constructing social orders either through stabilization or destabilization (Lash, 2000; Lupton, 1999; Rayner, 1992; Scott, 2000). Douglas and Wildavsky's concern for boundary maintenance gives a parallel political-anthropological narrative to Haraway's analysis of the immune system. However, whereas for Douglas and Wildavksy, the normativity of ordering deployed by the immune system is identical to that of the social system, Haraway is less adherent to such a naturalized logic and less favourable to the politics of propping up a centre that holds. That is to say, her preferred normativity is not necessarily that of immunization, of eliminating the different as pathological. More specifically, she argues that current developments in technoscience show that the ongoing convergence of biotechnologies and cyber-technologies suggests that there is a growing disparity between the 'old' (modern) normativity of immunization, and the emergent normativity of transgression.

In *Modest Witness* Haraway shows that contemporary technoscience is no longer exclusively concerned with forms of 'containment'. Instead, its institutional logic – which was already differentiated in terms of disciplines and paradigms – is scattered across a range of settings and practices. That is to say, in contrast to Beck's assumption of a monolithic science based on subpolitics, and more in line with Latour's assertion that technoscience is not any different from other systems in that it too relies on alliances, networks and cycles of credit to enhance strength, she points towards differences within technoscience that may themselves become nodal points in the formation of transformative political action. However, she also insists that the accumulation of credit within science is still predominantly coded by a logic of capital and governance that supports existing relations of domination. At best, technoscience produces pathologies (risks) that are dysfunctional, but they are still far away from becoming counter hegemonic. Indeed, pathologies are easily transformed into functionalities, as risks become opportunities, and are being valorized in terms of capital and governance.

However, what must be stressed above all is that Haraway's insistence on open, differential systems is politically motivated. This means that while she embraces ambivalence, she rejects nihilism and relativism. Neither politics nor objectivity are merely a matter of will to power. Indeed, what she refers to as partial objectivity is not necessarily and exclusively enforced by domination, but by forms of transgression of previously fixed boundaries and classifications of 'alterity' (difference/otherness). This could be understood as a critical form of reflexivity: a looking back upon the research process, the conditions of interpretation and tacit assumptions, as a way of problematizing one's own taking for granted. In this sense, situated knowledge can be seen as a form of appropriation without (total) incorporation. Haraway's critical reflexivity allows one to engage not only with the other-out-there, but also with the other-within, that which upsets the integrity of one's 'own' (appropriated) integrity.

The other must therefore not be idolized or fixed. The other does not hold a moral superiority. The other cannot be fully incorporated; but we are always invited to appropriate alterity. No 'given' can be taken for granted without some form of violation. The violation of the other is inevitable; the other is within; hence the violation of the self is inevitable. Partial knowledge is corrupt; it distorts. The spatio-temporality of cyber technologies has corrupted the social and violated the bodily integrity of the humanist subject. Humanism is itself a technological enframing, based on the denial of alterity and the imposition of the primacy of the idea of the human. But the other, the inhuman, is already within; it constitutes the human (Lyotard, 1991).

The corrupted human is the modest witness, a genetically manipulated information-processing device. It appropriates memories and traces of alterity that may otherwise be all too easily forgotten by the actualization of modern 'man's' will to power. What informs the epistemology of the modest witness is not less politically engaged than that of standpoint feminism (e.g. Harding, 1986), although it is not encoded in the binary-oppositional language of that particular epistemology. Haraway knows that we cannot simply oppose the techno-biopower of Science by universalizing a claim to essential womanhood; nor that we can simply erase the legacy of Christianity in the figuration of hyper-modern visual culture. Successful survival is possible only on negotiations between these multiple and incompatible legacies.

The critical reflexivity of situated knowledges returns in a new term: diffraction. Diffraction is a fragmentation of vision through the use of multiple foci and diffused light. It is a form of repetitious doubling that at once destabilizes the whole, by eradicating the primacy of the original form, and presents a poetic distortion of the technological enframing of visual technologies. It thereby intervenes in the otherwise linear narrative of origination and duplication that constitutes the main political drive of technological culture as the triple domination of 'man' over nature, divinity and history. Diffraction questions the assumption of original integrity which constitutes the politics of immunity and sovereignty. Diffraction makes it impossible to speak of 'an' identity; yet it still enables identification. It is a form of writing, of making sense, that intervenes in the technological enframing to distort its distortions and challenge its challenging-revealing. And it all simply begins with bearing witness.

Symbiosis and assemblage: Gilles Deleuze and Felix Guattari

Like ANT, Haraway's cyborg feminism maps the relationships between humans and technologies (and animals) and problematizes the boundaries between the human and non-human as permeable and unstable; open to playful modification and challenge, in a continuous process of becoming. In the previous chapter we have noted that (Latour's) ANT suffers from three major weaknesses: its lack of a historical conception of 'transgression' and radical systemic breakdown; its exclusive focus on 'presence' at the expense of absence; and its inadequate restriction of motivation to a 'will to power'.

Haraway's work offers a clear advance in terms of the first criticism. Her conceptual tool kit is radically historical in that it focuses on the central role of systemic transformations over time. In this sense it is well equipped to deal with the political dimension of risks and technological culture. By invoking both a notion of the pervasive impact of long-term historical legacies such as capitalism, patriarchy and governmentality and the more discontinuous ruptures that are effects of the contradictory nature of these forces, her understanding of the technocultural biopolitics of risks is well attuned to actual political struggles that currently exist.

In terms of the other two criticisms, however, some questions still remain unanswered. Although Haraway's work does open up a considerable scope for reconsidering technological culture as political, and thus to understanding risks as more than technological infidelities, her notion of the political – unfortunately – tends to remain too closely tied in with a metaphysics of presence. Absence seems to be a rather redundant category for cyborg feminism too, except perhaps in terms of potential. There is very little scope for theorizing 'the unconscious' in Haraway's work, simply because her understanding of politics is also too closely tied to intentions and reasons, rather than desires and subliminal forces.

However, we should resist the temptation of responding to this call to reflect on the unconscious by simply importing the dogmas of psychoanalysis. Rather, if we take Latour's principle of irreducibility seriously, we cannot embrace the unconscious in terms of narrative substitutions or even archetypes. That is, we must refrain from trying to read into absence any ego-defence mechanism such as denial, projection, substitution or sublimation. We would then simply reduce absence to what is already known (e.g. lack of phallus). Indeed, Haraway's concept of diffraction shows that women are more than female men or alter egos.

The first step in attempting to open biopolitics to an understanding of absence that is not simply a present enacted at a distance is therefore a critique of the Freudian unconscious. It is here that the work of Deleuze and Guattari is of seminal importance. In the *Anti-Oedipus*, Deleuze and Guattari (1983) have provided what can probably be called a final death blow to the psychoanalytic drive to reread into the unconscious a monolithic narrative of Oedipal desire. Psychoanalysis replaces the multiplicity of free associations with the identity of the proper name: 'It's daddy' (Deleuze and Guattari, 1988: 38). 'Freud tried to approach crowd phenomena from the point of view of the unconscious, but he did not see clearly, he did not see that the unconscious itself was fundamentally a crowd' (Deleuze and Guattari, 1988: 29). The concept of actor networks rightly posits that it is not the unity of the element (e.g. an immutable mobile), but their multiplicity that determines the successful sustenance of this ensemble of forces.

Unfortunately, this as such is still problematic for it emphasizes the primacy of stability over flux, position over flow. The notion of 'multiplicity' articulates first and foremost the transformative capacity of the unconscious as a crowd-becoming; a disposition of fixed interests. In other words, whereas Latour's network of irreducibilities maintains a sense of quantity and divisibility of the members and concentration and sociability of the whole, the Deleuzoguattarian notion of multiplicity is that of a small 'pack' of dispersed, non-decomposable, nomadic and peripheral metamorphoses (1988: 33).

Finally, the third weakness of ANT concerns a lack of concern over 'motivation'. It is here that in particular Latour himself becomes victim of reductionism. Everything returns to a will to power. Haraway's cyborg is indeed an advance because it is able to embrace a more diverse set of desires; however ultimately for the cyborg too, what is at stake is still survival. In this sense, Haraway's notion of the normativity of immunization still adheres to an implicit 'organic' or 'natural' logic, which assumes that the subject of immunization is able to differentiate itself from others and constructs boundaries accordingly. In this sense, immune systems operate through autopoiesis.

There is, however, a way out. Whereas in Haraway's earlier work, the concept of 'identification' – especially in her analysis of the immune system – implicitly corresponds with Maturana and Varela's postulate that all organic life is geared to autopoiesis, her concept of diffraction enables a break from this. In *Modest Witness*, the organic or natural logic of autopoiesis is being eclipsed. A *modest witness* is not exclusively concerned with survival. Diffraction actually entails a loss of that centripetal motivation. Instead, and for Haraway this is a historic event, we begin to witness the ascendance of a new logic of life, *symbiosis*. This shift is primarily due to historical developments in technoscience, war machines and commerce.

Symbiosis is a process by which two different genetic types merge to constitute a new one. It is a trans-genic process, a bit like the cyborg or Oncomouse^{TM}. The concept of symbiosis was first developed by Albert Bernhard Frank, a German botanist to describe the mutual interdependence between two distinct species. However, it was another German botanist, Anton de Bary, who subsequently developed the idea in relation to parasitism (Ryan, 1996). However, as until the 1960s mainstream biology was preoccupied with Darwin's evolutionist concept of natural selection through elimination rather than association, the concept of symbiosis had been largely neglected until the late 1970s. An important contribution in the revival of symbiosis theory has been the work of the biologist Lynn Margulis (1993). In particular with the publication of her *Symbiosis in Cell Evolution* in 1981, it became quickly apparent that symbiosis could explain far more effectively than natural selection how new organisms could 'emerge' (Sapp, 1994).

The commonly held definition of symbiosis is the one developed by Werner Reisser as 'the interaction not between different species of life forms but between dissimilar genomes' (cited in Ryan, 1996: 280). This definition thus incorporates viruses. Viruses are strictly speaking not life forms, but the concept of symbiosis is particularly insightful in the explanation of their semi-autonomous existence in relation to other life forms. Symbiosis, ultimately, results in the formation of a new life form. However, it rarely just 'happens' spontaneously but takes place in stages: predatory parasitism, benign parasitism, co-evolution and ultimately, the birth of a new, more complex, organism. Hence it is not incompatible with the concept of evolution (Wills, 1996). The main difference, however, is that rather than relying on discrete genomes entering into combat to settle the question of who is the fittest for survival, its perspective on evolutionary patterns entails much greater levels of complexity, volatility and transgression, in which information processing is not a matter of elimination and selection, but of convergence and transcoding (Ansell Pearson, 1997).

If we understand that the basic process in symbiosis is a form of interaction between two or more *different* information-processing systems, that in turn work to manipulate and modify their environment to better their chances of survival, then this should include all possible kinds of information-processing systems. What becomes furthermore apparent is that there are different intensities of symbiosis – varying between complete synthesis (transforming into a singular information-processing system, a new genome) to far more modest forms of nesting, interaction and cooperation. Finally, symbioses do not necessarily have to be successful. Even in biology, where the focus is predominantly on explaining successful symbioses, there are many more cases of 'failed' ones. Infections, disease and death are among a range of possible outcomes of failed symbioses in which one or more of the participants do not benefit from or are harmed by the association; they are cut off from further becomings.

Deleuze and Guattari use the notion of *machinic assemblage* to describe the particular constellations of forces (which include positions as well as flows, singularities as well as multiplicities) that constitute (hence partly institutionalize) as well as displace a transformative practice of becoming. In other words, rather than opposing actor networks and multiplicities, the notion of machinic assemblage can be used to preserve a sensitivity to movement and transgression in what would otherwise only be an account of institutionalization.

> There are no individual statements, only statement-producing machinic assemblages. We say that the assemblage is fundamentally libidinal and unconscious. It is the unconscious in person. For the moment we will note that assemblages have elements (or multiplicities) of several kinds: human, social and technical machines, organized modal machines, molecular machines with their particles of becoming-inhuman; Oedipal-apparatuses (*yes, of course there are Oedipal statements, many of them*); and counter-Oedipal apparatuses, variable in aspect and functioning.
>
> (Deleuze and Guattari, 1988: 36, original emphasis)

The notion of assemblage allows us to extend actor-network thinking to all kinds of practices of sense-making, and not exclusively those of official or 'royal' technoscience. It heightens a sensitivity for the between-forms, the hybrid bastardized offspring of molar and molecular multiplicities. It also enables us to conceptualize symbiosis beyond strictly biological discourse, without having to resort to simple metaphorical rhetoric. According to Deleuze and Guattari, assemblages

> operate in zones where milieus become decoded: they begin by extracting a *territory* from the milieus. Every assemblage is basically territorial. The first

concrete rule for assemblages is to discover what territoriality they envelop, for there always is one . . . The territory is made of decoded fragments of all kinds, which are borrowed from the milieus but then assume the value of 'properties' . . . The territory is more than the organism and the milieu, and the relation between the two, that is why the assemblage goes beyond mere 'behaviour'.

(Deleuze and Guattari, 1988: 503–4, original emphasis)

Rather than a complete auto-erotic event, the arrival of the assemblage is the result of symbiosis. Margulis (1993) focuses on bacteria, protozoa and other micro-organisms to argue that more complex cells evolved out of a symbiosis between less complex cells, often engaging in a combination of functions. All complex forms of life, plants, humans, animals, fungi and even single-nucleic cells are products of symbiosis – assemblages of different information-processing devices. In this sense, the autopoiesis of a more complex system is nothing but a gathering, an ensemble, of information-processing devices that decode the 'milieu' in which the emergent assemblage literally 'takes place' (becoming-territory) as a symbiont (Ansell Pearson, 1997: 134).

This means that the theory of symbiosis can be extended to interactions and associations between more complex life forms. Indeed, symbiosis has been an extremely useful concept in describing, for example, the relationship between aphids and ants. Ants feed off the sweet-sticky substances that are shed by aphids; in exchange for this, ants attack any predator that wishes to feed on the aphids. Likewise, one may argue, there is a symbiosis between humans and cattle. However, these relationships have all involved forms of cooperation that are benign and not aggressively parasitic. This contrasts, for example, with the symbiotic relationship between ants and rattan cane in Borneo, where the ants build their nest around a rattan cane plant to feed off the nectar produced by the cane but not to its destruction, while scaring off other herbivores with whom it competes over the plant (Ryan, 1996: 281–2). The plant suffers from the ants, but survives, while it would be more severely damaged by herbivores.

In all these cases, a symbiosis between different genomes has taken place. However, not all symbioses result in the formation of a new genome. The synthetic form of life, say rattan cane/ant, is not an integral and unified organism with a singular DNA, but remains an association of different parts. At first sight, this may also be seen as applying to the symbiotic relationships between organisms and viruses. Chimpanzees, for example, are thought to have a symbiotic relationship with the retrovirus HIV. This is not uncommon as there are, for example, around 10,000 retroviral traces in the human genome alone (Wills, 1996). Although chimps can be infected with HIV, this does not result in the development of AIDS. Chimpanzee and HIV simply coexist. Other primates, however, who may threaten to endanger the habitat of the chimpanzee might be infected by HIV. They could, as a result, develop AIDS, usually leading to their death. This makes a viral infection far less 'accidental' and much more intricately connected with a sort of self-regulating ecosystem (Ryan, 1996; also see Shope and Evans, 1993).

Hence, Haraway's cyborg could be understood as a symbiotic assemblage. In this sense, her work mirrors Deleuze and Guattari's concept closely. The main difference, however, is that for Haraway the shift from autopoiesis to symbiosis is a historical event. The arrival of intelligent machines has displaced the self-evident centredness of integrity and authenticity within technological culture, towards a far less humanist and more playful field of hybridities. This historicist postulate, however, is not accepted by Deleuze and Guattari, for whom the very principle of nomadology has always been symbiotic and operates through shifting iterant relationships with the more autopoietic state apparatus. There have always been symbiotic assemblages throughout history, that is because life itself is not qualitatively different from any other technical system (Ansell Pearson, 1997: 149).

More specifically, in contrast to Haraway, Deleuze and Guattari do not see the diffusion of technological culture as a historical phenomenon; they do not speak of a no-longer; instead their 'nomadological perspective' appropriates Nietzsche's notion of the 'eternal return'. It is not that humans and technologies were once but no-longer separate; humans and technologies are merely effects of forces – iterative transformations of energy and matter (Ansell Pearson, 1997). In other words, whereas Haraway perceives the question of value to be one that emerges out of the meeting between humans and technologies, Deleuze and Guattari perceive humans as incorporations of technological culture, parts of a wider machinic assemblage in which they perform their part. Value, therefore, is derived not from rationality or playful politics, but from the force generated by the assemblage itself.

The point to make here is that a Deleuzoguattarian understanding of the technocultural enframing of risk points towards the excessive flows of risks (as a multiplicity of desires and motivations) within the enframing itself – normativity is not externally imposed by some abstract rule ('pure reason'), but a force generated from within. In other words, whereas the technocultural understanding of risk generates a perception of risk as exterior to technoscience, this exteriority is itself the effect of the very same technoscience that cultivated the particular risk sensibility. If we take this point seriously, it means that the distinction between a closed (autopoietic) and an open (symbiotic) technoscience is far less straightforward than initially suggested. The self-referential (autopoiesis) and the other-referential (symbiosis) are mere vectors, existing in relative dependence. Indeed, if we understand technological culture in terms of 'flow', the fundamental opposition between the two disappears; instead there is only a notion of deviation-without-a-norm, forms of difference without identity.

> Every other difference, every difference that is not rooted in this way, is an unbounded, uncoordinated and inorganic difference: too large or too small, not only to be thought but to exist. Ceasing to be thought, difference is dissipated in non-being. From this, it is concluded that difference in itself

remains condemned and must atone or be redeemed under the auspices of a reason which renders it liveable and thinkable, and makes it the object of organic representation.

(Deleuze, 1968/1994: 262)

A nomadology of risk

A more specific discussion of Deleuze and Guattari's writings on the war machine and the state apparatus can be used to develop a sense of 'risk' in relation to 'violence' and 'transgression'. The state apparatus emerged in the form of a bounded territoriality, that was encoded in terms of law which is closely associated with a wider range of forces and techniques such as engineering, policing and military force. These were spatially distributed in (fortified) cities. Indeed, law is thus often seen as a preoccupation of the state apparatus.

Deleuze and Guattari stress that whereas we tend to understand law in terms of *logos*, which indeed suits the striated spatial organization favoured by the state apparatus, nomadology appropriates a different kind of law, that of *nomos*, which follows numerical and distributive modality of 'coding'. Law as *logos* is autopoietic, that is, it can deal with the operations of other systems only if they are being translated into the language of law. This makes law a specialization that – over time – tends to become inaccessible to other kinds of discourse. The *logos* of law is binary, either something is or it is not. According to Teubner (1989) this kind of law internalizes its own environment, attempts to become fully self-referential, and succeeds in that when law becomes sovereign (see Chapter 2). The main problem with this becoming-autopoietic is the core of the risk society thesis. It is the failure of communication, e.g. law that ceases to translate itself into other operating systems or technoscience that fails to engage a politics of representation. Once closed off from intersystemic communication, they can then perform a coordinating role only through mutual observation.

However, following Deleuze and Guattari, *nomos* operates on the basis of modalities of encoding and decoding. Its Greek origins refer to forms of enclosure, such as walls and fences. The aim of *nomos* is to regulate every flow, of whatever matter. *Nomos* itself belongs to no system in particular, but works as a 'code' or 'cipher' between systems. Its purpose is not to close itself off from any other operating system but to dissimulate itself throughout every aspect of information processing, enabling systems to continue to differentiate themselves while being connected. *Nomos* thus operates on the basis of the translation of assemblages by establishing equivalences between different singularities. The aim of translation is not to make everything the same, as in *logos*, but to enable infinite and continuous differentiation, i.e. *differance*.

The difference between *logos* and *nomos* also affects modes of warfare. *Logos* inaugurates order, hierarchical organization and the development of strategies. It informs the military apparatus, with its hierarchical authority structures and regimentations. *Nomos* writes itself in the formation of packs, who innovate but do not settle. It informs what is now termed 'international terrorism' which consists of spatially dispersed, largely autonomous ephemeral cells that disappear from the surveillance systems of the state apparatus exactly because they are so iterant. *Nomos* enables pure movement and deterritorializes, *logos* sets up a grid and territorializes domains. It is therefore extremely difficult to confront *nomos* with *logos*. This becomes tragically clear in the way in which the USA and its allies attempt to combat 'international terrorism' using military force. The ballistics of the state apparatus are so crude and unsophisticated that they destroy indiscriminately. Nomadic war machines, however, are far more precise in their targeting, even if they may take many casualties on the way, these are rarely seen as 'collateral damage'.

For *logos*, risk is primarily defined in terms of a lack of ordering. Risk is that which escapes the gridlock of organized, identified, territoriality. It exposes gaps in the defensive walls, in the security systems, and in the sealed logic of autopoietic law. For *nomos*, risk relates to a lack of mobility and movement, to incommunicability of information flows, to codes that cannot be decrypted. In the *logos* of warfare (i.e. the state apparatus), risks are exteriorities projected onto the interior; they are always from elsewhere, they threaten the boundaries of the centripetal organized system. In contrast, for the war-machine risks are interiorities projected outwards; they threaten to erect boundaries in the form of designated paths, walls and striations that stop centrifugal flows and lines of flight. Risks are ruptures in the assemblage, fragmentations of the pack, de-differentiations of codes.

When we translate this to technological culture, we can see that in modern society, risks tend to be conceptualized through logocentric systems of interpretation. That is, the majority of risk articulations take the form of boundary transgressions (Douglas and Wildavsky, 1982). Hence, the attack on the World Trade Center was branded a 'criminal act of war' by American officials. This exactly stresses the limitations of logocentric reason to come to terms with nomadology. However, in the same modern capitalist society, we see a completely different understanding of risk at work as well; for example the risks of stagnation and recession in the world economy, or the risks of inertia, lack of investment, lack of innovation, boredom, sameness and self-enclosures that prevent the formation of collective morality (Durkheim would call this anomie). Indeed, ever since Andy Warhol's perversion of modern art (Jameson, 1994), technological culture seems to have turned more in favour of the nomadic. There are now dozens of Deleuzean graduates and postgraduates flocking onto the labour market and into the academy, governance, commerce and the cultural industries. Indeed, *nomos* is returning to technological culture as nomads, tribes, packs, posses and crews are popular expressions of collectivities of youth today (Maffesoli, 1996).

However, we should be very careful extrapolating from these trends how technological culture works. There is more to technological culture than fashion design. Technological culture embodied by the institutions of technoscience and governmentality are still very much dominated by *logos*, and so too are law and the military. The media may seem to be more evenly balanced but are far more complicit with the state apparatus than careless scanning of incidental writings by freelance journalists in a few left-wing newspapers might suggest (Tulloch, 1993). This leaves us with commerce, which is indeed perhaps the most nomadic of modern institutions, if only because it is driven by the relentless deterritorializing force of capitalism. It thus becomes a strange and uncanny realization that the most nomadological institutional force in modern society is engendered by capitalism.

The problem here is that if we limit ourselves to institutions, it is obvious that we will not find a lot of *nomos*, this is because modern institutions are by definition logocentric. Not only do they work on the basis of logocentric principles, but also they elaborate and enhance them. Nomadism comes from the outside, from the fissures and cracks in the institutionalized disorder of everyday life, from small gatherings of people, emerging in particular events (community), but easily dissolved into anonymous citizenry when the police arrive. Nomadism travels through the cracks in the virtual walls of institutionalized modernity that are called 'risks' in logocentric modes of organization. These risks enable nomads to make connections with certain parts of the state apparatus, to form new assemblages with them, on the margins of social order. Hence, various gay and lesbian movements were able to travel alongside highly nomadic HIV, mobilizing public concern for this dreadful disease, and as a result accumulate both financial, social, political, symbolic and cultural capital within the newly formed networks of AIDS concern (e.g. Oppenheimer, 1997; Paillard, 1998). Nearly all established institutions of modern society were infiltrated and forced to change their innate inertia and apathy into a more adequate response to the threatening pandemic (Mann *et al.*, 1992). New forms of disseminating information were introduced. Old debates, for example on the moral virtues of barrier methods of contraception (especially the condom) suddenly seem to have become redundant. Family planning is no longer optional; it is a must. Even after 20 years of scientific research, HIV still makes more sense as *nomos* than as *logos*. Its objectivity has still not been stabilized; as a result it does not allow itself to be subjugated to any binary logic; ambivalence remains a more powerful code (Ironstone-Catteral, 2000).

HIV can tell us a lot about the logic of the risk society. The risk society has emerged because of a failure of logocentrism and its correlate, the metaphysics of presence. Before BSE, Oncomouse^{TM}, Chernobyl or HIV, there were of course already many risks and they too were embraced by moral panics, mobilizing public concern and specific political responses. However, these concerns and responses addressed a set of systems, often headed by technoscience, and to a lesser degree governance and law, that were expected to deal with them if not swiftly then efficiently. The risks were received in terms of a *not yet*, but were never projected onto an infinite horizon. In the 1960s and 1970s, most people believed that scientists would find a cure for cancer, that ecological degradation was reversible and that one day humankind would be delivered from all infectious diseases through vaccination and antibiotics. Since the 1980s, scepticism has increased and faith in technoscientific fixations has decreased dramatically. The old assemblages of risk management have become dysfunctional and as a result, the risks of risk management have become more and more apparent, adding to an already proliferating risk sensibility. This, in a nutshell, is Beck's risk society thesis.

However, by invoking an opposition between *logos* and *nomos*, we can see that the very meaning of risks has been destabilized with the erosion of the hegemony of the state apparatus and its logocentric mode of organization. We can also see why it is that what are risks for some may be opportunities for others. Moreover, we can explain the formation of new assemblages, mobilized around new articulations of risk perceptions. Hence, invoking this distinction enables us to stress that the notion of technological culture as an enframing of risk needs to be steered away from the opposition of good versus bad, and instead, moved towards a sensitivity to a biopolitics of 'motivation'.

The key opposition in motivation can be understood as being between preservation versus transcendence. When one links 'motivation' with biophilosophical concepts such as 'excess', 'force' and 'flow', it can be argued that risks are not simply ordered and revealed in technological culture, but that technological culture is simultaneously being ordered and revealed by risks. Logocentric systems are geared towards autopoiesis. They follow the basic double bureaucratic logic of ordering means-to-ends before turning means into ends. Beck's risk society is run by aspiring autopoietic systems that fail to converge around dealing with pressing issues with catastrophic potential (such as nuclear or genetic risks). However, it would be false to suggest that these systems simply fail to communicate. The point is merely that systems geared towards self-preservation often encounter many obstacles to inaugurating new assemblages, and developing innovative modes of response.

Innovations often have to come from the nomadic forces of commerce and communal and social movements. It is important to stress, however, that these elements are not completely absent in the risk society. Indeed, they are so apparent, that Douglas and Wildavsky (1982) attribute to social movements a role of risk entrepreneurship; using the language of risk and blame to destabilize the centre (Lash, 2000). However, using a bit of Deleuze and Guattari's *Treatise on Nomadology*, we could turn the tables on Douglas and Wildavsky, by arguing that these nomadic forces may hold the key to supporting a centre that holds.

Contingency, risk and the monad: Jean-François Lyotard's apocalypse

Such an observation brings us, finally, to the work of Lyotard (1991) on the inhuman. Lyotard asserts that the movement within technoculture is nothing but an attempt to survive the incumbent collapse of organic life on earth. He notes that technoscience attempts to develop thought without a body, hence perfecting the millennial Platonic ideal. His notion of the inhuman, as that difference within that constitutes our very human condition, is very similar to Deleuze and Guattari's notion of the machinic assemblage. It is the becoming-human of technology that Lyotard postulates as being the essence of the Inhuman.

> A human, in short, is a living organization that is not only complex but, so to speak, replex. It can grasp itself as a medium (as in medicine) or as an organ (as in goal-directed activity) or as an object (as in thought . . .). It can even abstract itself from itself and only take into account its rules of processing, as in logic and mathematics.
>
> (Lyotard, 1991: 12–13)

In asking the question 'can thought exist without a body?', Lyotard seeks to rethink the *Ur-Erde*, i.e. the first grounding principle, of being and thinking. Similar to Deleuze and Guattari, he asserts that it is multiplicity (the differend) that constitutes the typical but transitory arrangement of matter-energy that we call organic life. It is this difference (that manifests itself, for example, in terms of 'gender' or the opposition between thought and body) that causes infinite thought. Hence rather than locating technoscience inside the psycho-physiological hardware of the body (alternatively substituted with nature, reality or matter), it is a 'cosmic circumstance'.

> As a material ensemble, the human body hinders the separability of this intelligence, hinders its exile and therefore survival. But at the same time the body, our phenomenological, mortal, perceiving body is the only available *analogon* for thinking a certain complexity of thought.
>
> (Lyotard, 1991: 22)

In other words, machinic assemblages are not the product of wilful intentional *human* choreographies, but inscriptions of the monadic drive of technoscience to exclude contingencies and secure *a* future. Just as all the other actors in actor networks, humans are not to be seen 'as an origin or as a result, but as trans-formers ensuring through technoscience, arts, economic development, cultures, and the new memorization they involve, a supplement of complexity in the universe' (Lyotard, 1991: 45).

Hence, Lyotard's particular displacement of the humanist imperative of equat-ing action with consciousness, which is central to the metaphysics of presence, allows us to complete the move from an individual, molar, mode of rationality, to one which is cultivated through and as multiplicity. Instead of deliberate strat-egy, Lyotard locates 'technology' as a 'spatialization of meaning', in the trans-formative processes (modes of becoming) of 'breaching' (disrupting the formation of habits on the basis of regularity or repetition), 'scanning' (repeti-tious 'remembering' or experimentation as a foreclosure of ends) and 'passing' (working-through, anamnesis).[4] Acting at a distance, the enrolment of absence then becomes a practice of telegraphy. As we know from Derrida (1978), writing (*écriture*) refers to an inscription of traces (*graphes*), that are neither present nor absent, but are the setting-into-work of difference, i.e. temporization and spatialization (Van Loon, 1996a).

However, Lyotard's problematic is, unlike Deleuze and Guattari's, not that of becoming but that of judgement, or better, heteronomy (Curtis, 2001). Lyotard's inhuman is rather similar to Beck's paradoxical figure that tries to overcome the delegitimation of technoscience by delegitimating technoscience. The inhuman is thus a tragic figure whose success is the failure to succeed. However, Lyotard's problematic allows us to turn to the question of ethics more directly. For him, ethics would be a movement within technological culture against technocultural ordering and revealing – a movement that works through recollection, remembering and rewriting the very frames that revealed and then concealed the risks as virulent abjects within the machinic assemblage.

The concept of 'virulent abject' is an appropriation of Kristeva's notion of the abject in conjunction with Deleuze and Guattari's concept of the nomadic. Lyotard's argument, that technological culture is taking the form of a monad, which exists on the basis of programming its own destiny by incorporating 'contingency', presents a powerful image of a modernism that refuses to surrender. Indeed, a superficial scan of contemporary global politics and commerce immediately makes clear that those who have been allocated the responsibility of keeping the system going despite its various trials and trepidations may have lost their meta-narratives, they have not lost their performativity. Whatever happens, managers and politicians are still driven by a desire to plan and control, to produce maximum effects with a maximum of efficiency, against a minimum of costs. In order to do that, more and more aspects of our being-in-the-world have to be rendered transparent and open to treatment. Secrecy and privacy are breeding grounds of risks that must be 'challenged forth' and revealed, in order to be managed.

Lyotard shows, however, that whereas preservation is the principal motivation of monadic ethics, the effect of it is quite the opposite. To illustrate this, Lyotard tells a fable of the monad that needs to transcend its current dependence on embodiment if it is to survive the collapse of the Sun in 4.5 billion years. Risks can be seen as signals of contingencies that are yet to be incorporated. Yet, by their very nature, they are also elements of transgression. Once incorporated, they start generating new contingencies that reveal themselves as risks that need to be incorporated. Pestilence, virulence and contagion therefore mark the becoming of the monad.

Lyotard's apocalyptic narrative reflects the iterant historical struggles between *logos* and *nomos*, between autopoiesis and symbiosis and between the biopolitics of immunity and the biopolitics of diffraction. At this crucial point in time, as the modern world desperately hangs on to logocentrism, to the neat orderings and classifications informed by technological culture, the nomadological nature of risk is indeed more prominent. This explains the close associations between risk, contingency, insecurity, irregularity and pathology. This explains why containment and immunization prevail, at the expense of transgression and innovation. However, despite the force mounted against risk, the fate of logocentric state apparatuses remains tragically connected to risk. Modern technological culture is thus driven by a paradoxical thirst for annihilation in the name of its own survival.

5   A theoretical framework

Risk as the critical limit of technological culture

The purpose of this chapter is to provide a synthesis of the main points of the previous three chapters to provide a theoretical answer to the question raised in Chapter 1: What is the specific relationship between risks and technological culture and what are the consequences of this relationship for the organization of 'the social'? Its objective is to construct a theoretical framework for the analysis of specific risks in 'our' modern, western technological culture. This is a prerequisite for understanding how risks have been handled by some of the institutions that reproduce this technological culture, which is the objective of Part II.

The chosen starting point was Beck's rather historicist account of how the risk society signals a transformation of modernity. The aim was to show that this change has strong cultural implications that deeply affect how we live with technology. Beck's thesis evolves around a paradoxical juxtapositioning of a number of key sociological concepts: individual and society, goods and bads, risks and hazards, uncertainty and security, subpolitics and democracy, scarcity and excess, relation of production and relations of risk definition, nature and culture, reality and representation, etc. The main aim of these dual oppositions for Beck was not to sketch the move from one to another, but to show how they operate in relationships that are themselves changing.

Whereas Beck's work indeed provides a major contribution to thinking about the 'third terms' which would dissolve these oppositions, more work needed to be done to develop it into a theory that is at once empirically informed (for example about science and technology), politically radical (otherwise it is too easily co-opted by existing structures of domination) and conceptually innovative (to develop alternative concepts to those of the aforementioned dualisms). Rather than a disqualification of the risk society thesis, it is a matter of extending its theoretical scope in a dimension that has thus far received less attention in the sociology of risk.

A first step was a turn to Latour and Actor Network Theory to explore in greater detail how technoscience actually deals with risks. ANT allows us to consider risks as virtual objects that are 'fixed' in alliances between actors (humans, technologies, spirits). Risks are modes of enrolment that have to be visualized, signified and valorized in cycles of credit which constitute the basic structure of technoscientific work. However, whereas Beck's work may be seen as unduly historicist, Latour has a tendency to be too dogmatically anti-historicist and to reduce historical changes to equivalent social processes. His understanding of actor networks is therefore ill equipped to deal with the kind of notions of 'turning' and 'transformation' that the risk society thesis seeks to endorse. Whereas it is sensible to question whether the apocalyptic scenario of the risk society is not merely another strategy of enrolment, it is equally wrong to rule out that a significant turning has occurred, simply by referring to an (alleged) lack of empirical verification; the glass is neither just half full nor just half empty but always both. Latour's work risks sliding into a metaphysics of presence and undervaluing instability and ambivalence.

To counteract on those weaknesses, Chapter 4 turned to biophilosophy which – especially in the works of Nietzsche and Deleuze and Guattari – has had a major impact on Latour's thinking. From the amorphous and sprawling 'genre' of biophilosophies, I restricted myself to discussing the works of Haraway, Deleuze and Guattari and Lyotard to construct a little narrative based on a politics and ethics of difference. From Haraway's writings on immune systems, cyborgs and situated partial knowledge we obtained a subtle vision of the way in which technology constitutes who we are and how we perceive ourselves. Her work is radically reflexive, most clearly expressed in the concept of diffraction as the loss of integral unified perspective. Deleuze and Guattari enabled a continuation of this theme, but also a radicalization of the notion of difference as a perpetual force of vitality and movement. Their concept of assemblage as a 'difference engine' presents a most emphatic attempt to think through the concept of life as continuous assembling of singularity and multi-plicity. Lyotard's more sceptical reinterpretation of technological culture as the work of the monad to create the possibility of thought without a body gave a more pessimistic and apocalyptic twist to the biophilosophical narrative, which is crucial if one wants to understand risk as a limit of technological culture.

This chapter attempts to argue in more detail that the risk society itself provides a concept of 'turning' within technological culture. Appropriating (rather heretically) some ideas mainly from Heidegger on technology, we will see how it is possible to understand risks as far more fundamental to the undoing of modern society than is often assumed by critics of Beck. The notion of risk as a virtual object will now be coupled with the concept of assemblage to show how the becoming-real of risks entails a process of translation involving visualization, signification and valorization. This enables us to argue that technological culture is based on 'structurally induced needs' (Lodziak, 1995) in which our lives are organized to provide technological input – usually in the form of time and energy – through which technical systems are able to sustain themselves as if they were autonomous, living, entities. The aim of this is to argue that from within, from without, and against technological culture, new and old apocalyptic sensibilities have (re-)emerged that have appropriated risks to undermine this autopoietic movement of technical systems and reveal new openings for radical change.

The logic of en-presenting: visualization, signification and objectification

> What has the essence of technology to do with revealing? The answer: everything. For every bringing-forth is grounded in revealing . . . The possibility of all productive manufacturing lies in revealing. Technology is therefore no mere means. Technology is a way of revealing.
>
> (Heidegger, 1977: 12)

If one wants to develop a critical understanding of the role of technology in modern through, two of Martin Heidegger's (1977) seminal essays 'The Question Concerning Technology' and 'The Turning' could be useful starting points. What connects these two essays is Heidegger's ongoing concern with Being, or more specifically, how the question of Being has disappeared in modern thought as a continuation of what he referred to as 'the metaphysical tradition of Western philosophy'. Even more specifically, the essays turn to the notion of technology as a culmination of modern thought, a mode of being in which modernity reveals and conceals itself most fully.

Heidegger starts his exploration with two commonsensical understandings of technology as means to an end and human activity. Both constitute an instrumental understanding of technology, as being in service of humanity. Whereas he wants to maintain the (descriptive) instrumentalist ethos of technology, the essence of technology for Heidegger is neither its toolness nor that it is a piece of equipment: instead, it is a way of making present, or *en-presenting*. The notion of en-presenting (*Gegenwärtigen*) is derived from Heidegger's (1927/1986: 326) *Sein und Zeit* (*Being and Time*). It covers both 'making present' and 'bringing into presence' (also see Blattner, 1992: 117; Van Loon, 1996a: 76). It relates to, for example, en-countering, en-gaging and en-abling. These are all transitive verbs which strongly emphasize relational processes. The other German verb for presenting Heidegger invokes is *Anwesen* (translated as being present), which is intransitive and non-relational. Moreover, the prefix 'en' highlights forms of action that are part of a creative process (*poiesis*). En-presenting is different from presenting in that it never simply delivers what is 'already there' but creates something. It is an attribution of presence and thus never simply a spatial encounter of immediacy but also implies temporality; it takes time. The fact that in English we do not have a word for *Gegenwärtigen*,[1] and in German the word is more commonly used as noun, i.e. *Gegenwart* (present), signals that our language is indeed deeply imbued in a metaphysics of presence in which being is fully conflated to the immediacy of the utterance (Derrida, 1982; Van Loon, 1996a). The metaphysics of presence suggests that there is no difference between Being and Language; i.e. the utterance immediately brings into presence its own referent. We thus find it difficult to 'speak' through the interval of 'coming in to being', the temporality of being constantly alludes us in our desire to fix an unmediated presence in language.[2]

En-presenting is a form of disclosure. In *Sein und Zeit*, Heidegger discussed its basic structure in terms of a triad: so-foundness (*Befindlichkeit*), telling (*Rede*) and understanding (*Verstehen*) (also see Haugeland, 1992: 27). This triad is quite similar to one already alluded to in previous chapters in terms of visualization, signification and valorization. Using Heidegger, it is now possible to deepen our understanding of these three terms.

En-presenting is first of all a form of encountering, showing, disclosure or making visible, that is, visualization. Visualization reveals what has been hidden. Heidegger uses the term *Entbergen* (revealing) to argue that the essence of technology is a bringing-forth (one could even think of creation – *poiesis* – here) of that which is hidden. That is, technology brings concealment (*Bergen*) forth into unconcealment. What is being revealed is not simply the matter of technology itself (*causa materialis*), but also its form (*causa formalis*), its end or *telos* (*causa finalis*) and what brings about the effects (*causa efficiens*) (Heidegger, 1977: 6). The character that binds this fourfold causality is revealing or bringing-forth (*poiesis*); all four belong together in this. The argument of this book is that problems start, i.e. that risks emerge, when these four aspects of revealing do not add up.

Technology is thus not a thing but a process or more precisely *flow* which includes matter, form, purpose and effect. Technology thereby shows us something: it discloses a specific trajectory of a particular matter, through its formation in production, its purposeful utility in action, but also its consequences, both manifest and latent. This revealing or unconcealment is the first modality of en-presenting. In Chapter 3, using Latour's distinction between presentation, representation and re-presentation, we have already discussed three different forms of en-presenting, all belonging to a different 'realm' or 'field' of knowledge: art, science and religion.[3] In all three realms, the meaning of 'presence' differs. Yet it is their juxtapositioning that enables one form to dominate and incorporate others. Returning to Heidegger, we can now argue that in the modern era, the dominant form is 'representation' and it is primarily produced by technoscience. Hence, the en-presenting that is the work of technology is primarily a form of representation.

In Chapter 4 we discussed Haraway and used her argument that modern technoscience privileges the visual. Consequently, we can also argue that the revealing of technology primarily takes a visual modality (Denzin, 1995). Visual representation, e.g. one that enables 'seeing reality' through a microscope (Hacking, 1983), is indeed a very powerful force in the technological conquest of 'man' over nature, divinity and history. Likewise in law, visual evidence, provided by eyewitnesses or surveillance cameras, equally plays a powerful role, as we can see in the discussion of the Rodney King videotape in Chapter 9 where seeing provided direct access to truth.

The dominance of technoscience is perhaps nowhere more apparent than in the concept of the risk society. Beck asserts that knowledge of risks is generated by expert systems and that perception heavily depends on technoscience. Whether they are recognized forms of technoscientific expertise or based on so-called 'lay knowledge', to be able to articulate risks in a technological culture one must have access to specialist knowledge. The emphasis is on expert systems rather than experts because of the inherently systemic nature of knowledge governed by the episteme of modern thought. These expert systems may be technoscience, financial institutions such as banks and insurance companies, managers, farmers or social movements; what they do is generating data. With these data they enable the visualization of latent and extrasensory pathologies. In all of these different forms of 'expertise', technology plays a major and dominant role.

However, visualization in itself is not sufficient. 'Raw' data are meaningless unless they are interpreted. Apart from disclosure, risk perceptions also require signification. Signification is the transformation of information into a meaningful utterance; it is the en-gendering of meaning. Whereas initial signification is also the domain of expert systems, particularly technoscience, the wider social articulation ('publication') is primarily the work of governance and media systems (especially journals and books, but extending to newspapers, magazines, broadcasting, internet and even more specific forms of dissemination such as leaflets and political lobbies). What they do is to attribute 'significance'; they determine 'what matters'. What is being disclosed in signification is the encoding of expertise in terms of 'common sense'. For example, the demonstrations performed by Pasteur were intended to convince not only the medical but also wider political fields of the significance of his germ-theory (Latour, 1988). Only a few among us are expected to read medical journals or specialist publications on econometrics. It is therefore of paramount importance that these sources are being further translated by other specialist apparatuses and organized forces (such as insurance agencies, stockbrokers and bank managers) into political discourse for the public sphere.

Signification – the entering into a symbolic order – always closely follows visualization for nothing is merely revealed as itself; it always-already engages a semiotic process for what is being revealed to become meaningful. A microscope, for example, is never a simply 'optic technology', it always-already implies both 'practices' of handling (Hacking, 1983) and reasoning (Knorr-Cetina, 1983). Although this is as such inevitable, what is specific about modern technology is that this semiotic process pulls the revealing process into a highly formalized and instrumentalist mode of signification, one which is dominated, as if it were, by *causa finalis* (utility) and *causa efficiens* (effectiveness). This thus immediately conceals the revealing by repositioning it into a pre-determined and fixed order of formalized statements and expressions under the heading of Science.

The consequence of this is that in modern society, technology has intensified the force and speed of revealing to a degree that it becomes simultaneous with 'ordering' (Winner, 1977; Pacey, 1983). As has been mentioned in previous chapters, the nature of modernity is a threefold ascendance of domination of 'man' over nature, divinity and history. As a result, nature, divinity and history become resources for modern 'man'. As they are being revealed by technology, they become replaced by a technological flow of matter–form–purpose–effect, and turned into resources. Not only nature but also divinity and history are being challenged and themselves turned into parts of the technological flow. Heidegger, referring only to nature, calls this *herausfordern*, which also refers to extracting, provoking or expedition. Nature thereby becomes a standing-reserve (*Bestand*) for technology (Heidegger, 1977: 14). For example, whereas agriculture used to take place in an ethos of care/concern/correspondence with the soil (cultivation), it is now a mechanized and (genetically) engineered food industry. This, however, easily leads to a romantic reductionism that assumes that once upon a time, 'man' lived in symbiotic homeostasis with nature. This, however, is not only untrue but also unnecessary. All that matters is that pre-modern 'man' simply had no other option but to interpret nature as an externalized force – like God and fate – 'he' did not have the means to reorder the processes upon which 'his' farming technology relied. In the modern world, however, technology as expedition has become an unlocking and an exposure. Nature is predisposed – ready to hand – at our command. *Bestand* has an even more profound meaning in the world of ICTs, where it relates to stocks of data files piled up onto machines, to be revealed through simple operations such as automatic searching and browsing.

> Unlocking, transforming, storing, distributing and switching about are ways of revealing. But the revealing never simply comes to an end. Neither does it run off into the indeterminate. The revealing reveals to itself its own manifold interlocking paths, through regulating their course. This regulating itself is, for its part, everywhere secured. Regulating and securing become the chief characteristics of the challenging revealing.
>
> (Heidegger, 1977: 16)

That Heidegger used the terms regulating and securing as the chief characteristics of modern technology is telling. It is important to recall here Beck's insistence on the risk society as moving beyond the insurance principle. What happens in the risk society is that the difference between ordering and revealing is becoming once again more visible. That is to say, whereas modern technology shows that revealing is also a way of ordering, the risk society reveals a paradoxical limit to that ordering: one in which the regulating and securing no longer seem sufficient. The question is whether this will change the nature of our technological culture.

Heidegger, writing about modern technology before the advent of the risk society, argues that by setting upon an object, technology places it within its own course of action. In other words, technology 'enframes' the world. It does so not by simply placing it within a framework (that is, similar to putting a frame around a painting), but by assembling, calling forth, ordering for use. 'Man himself' also becomes part of the standing-reserve, as human resource management so clearly teaches us. Hence, Heidegger argues that

> Modern technology as an ordering revealing is, then, no merely human doing ... The challenging gathers man into ordering. This gathering concentrates man upon ordering the real as standing reserve.
>
> (Heidegger, 1977: 19)

The challenging calls forth 'modern man' as an instrument of technology's own unfolding. This is not an assertion of technological determinism, but the logical implication of enframing as ordering-revealing. It is not beyond human activity, indeed the enframing is all 'done' by humans themselves, but it is neither 'exclusively *in* man, [n]or decisively *through* man' (Heidegger, 1977: 24).

'Enframing means that way of revealing which holds sway in the essence of technology and which is itself nothing technological' (Heidegger, 1977: 20). In order to avoid the all-too-seductive association between enframing and an imposed structure, it is important to always connect it with revealing. In a way, the double sense in which the verb 'to order' could refer to either 'to put in place' and 'to command' may work as a reminder that every form of enframing, or indeed of 'fixing' (including fixations), is more than an instrumental activity that 'man' imposes upon 'his' world. Whereas we are initiated first by the instrumentality of technology, its toolness, it is the essence of technology (enframing/revealing) which is concealed to the last. In this way, Heidegger speaks about enframing as a form of destining. What is being revealed therefore is an engagement of the future into the present. The future, the unknown, is destined by enframing/revealing as the danger – that which must be revealed. The danger of enframing is that in the unconcealment of that which it brings forth, it conceals revealing itself (Heidegger, 1977: 27).

This is why we could argue that in his own way, Heidegger was already anticipating the advent of a risk society. The danger of enframing is very much lived in a technological culture. It is embodied in a lack of care we display on a daily basis towards our environment, including each other. It is found in the uncritical acceptance of various compromises to the way we live that are forced upon us by technology. For example, the closure of local shops in favour of large hypermarkets forces us to be dependent on cars which forces us to generate extra income to buy a car, the energy it consumes and its maintenance. These needs have been structurally induced (Lodziak, 1995) by a challenging-revealing of modern technology in terms of convenience, comfort and efficiency. However, such needs are effective only once they have been culturally incorporated into habits, routines, ways of life, perceptions, sensibilities, yes even ideologies and the unconscious. This is why we can speak of a technological culture. It is a culture in which technology is inhabited and taken for granted, because it cultivates, constitutes, orders and reveals, but thus also conceals, our being human.

This transition from what is hidden into what is made present – which is the essence of technology – is not without consequences. As has already been noted, it entails a second movement in which the technology that reveals also conceals itself through signification (becoming-significant); that is, it turns the en-presenting itself into a mythical obviousness (a metaphysics of presence). An example of this is the way in which technologies involved in the revelation of genes or viruses have quickly disappeared as necessary parts of understanding what genes and viruses are; the latter have become 'taken for granted'. Their presence is mythically projected back onto the time of their concealment as if they were always-already 'there'. That is to say, in becoming effective, technology withdraws itself from the revealing, and thereby takes the form of 'obviousness' that we take for granted when 'using' specific technologies (see Chapter 1). This is the essence of technological culture.

As far as en-presenting requires entry into the symbolic order, hence an enframing by and through language, this withdrawal is inevitable. To a certain extent, one must forget 'how' to use a tool in order to be able to use it. Signification is a necessary part of being-in-the-world. Language is always before us. The interesting thing about signification is that it contains a double act. Signification, which in essence refers to the becoming-significant or the attribution of significance, refers to both a semiotic and a political ordering of sense. In the first mode, what is significant is a symbolic property, in the second what is significant becomes the product of the assertion of force, authority, violence, power, in short 'a will'. The point is not that enframing entails semiotically and politically motivated distortions, but that some distortions have taken on a more permanent and systematic character, and as a result have all too easily been taken-for-granted.

The political nature of technology immediately raises questions about the way in which we understand its force. As many theorists (e.g. Pacey, 1983; Vig, 1988; Winner, 1977) have already observed, the political and organizational embedding of technological processes is itself often hidden by a governmentality that seeks to neutralize its significations. This is why Feenberg (1991) is absolutely right in stressing the need for a more critical understanding of technology against both substantialist and instrumentalist theories that both project technologies as autonomous from the political.

For Heidegger, the taking-for-granted is the concealment of unconcealment. However, through Beck's concept of the risk society as beyond the insurance principle, we may be witnessing a peculiar turning, where the danger becomes the saving power. The essence of technology as revealing is never more clear than when technology becomes dysfunctional. For example, our collective dependence on the car was made extremely visible during the 'fuel crisis' that swept across Europe in September and October 2000. Suddenly, the economy ground to a halt; services could not function properly; people were left at risk; hypermarkets could not replenish their stocks, etc. Many of us may have experienced a crash of computer systems at work, which suddenly reveals how dependent we are on information processing. Because in the risk society, we can no longer simply trust expert systems and we are forced into taking responsibility for our own actions as well as those of others, we have begun to *anticipate* breakdown. The aftermath of the collapse of the World Trade Center on 11 September 2001 is in this respect a global historical anchor of this realization. It has itself been described as a historical turning point that evolves around a notion of risk linked to 'international terrorism'. Indeed, living with risk has become an anticipation of 'bad things happening'. Risk-as-anticipation is the turning point for modern technology as it has to embrace a future as an extended-presence (Nowotny, 1994), that is far less speculative and careless than one concealed in orderings based on regulating and securing. This would imply a major breakdown of our technological culture which has thus far embraced an uncritical stance towards enframing (Farvar and Milton, 1973; Vig, 1988). The turning, then, is a rupture in this smooth relationship between concealment and unconcealment. This is where regulating and securing can no longer circumvent the danger. This is when technological culture no longer 'works' to incorporate, interpret and support (understand) structurally induced needs.

However, there are also considerable and structurally endemic limits to the possibilities of actualizing such a turning. This is because even if we are able to 'see' and 'reason' (speak of), the danger, we might not be able to 'understand' it. That is, we might not be able to consistently support (literally understanding as in 'propping up') a turning by means of our own political action. The choreography of concerted collective action is not simply a matter of reading a well-written script (e.g. the Communist Manifesto); its interpretation has to go beyond *logos*, to enter into – for example – practical flows of gifts and exchanges. That is, whereas it is possible to reclaim 'a' street for a certain period of time with or without the consent of authorities, it has proven a lot more difficult to reclaim 'the' street on a more permanent basis. This is because to do that, one needs to organize a collective will that can be sustained by institutions of law, governance, commerce, finance and – if necessary – military force (Pacey, 1983).

Indeed, apart from visualization and signification, there is a third dimension at work in en-presenting. As has been said, the attribution of significance to what has been visualized is both a semiotic and a political process. However, for this significance to affect anything, value needs to be attributed to it. That is to say, it must be valorized. Valorization enables a transformation of flows; it establishes connections between nodes in networks and enables operators to communicate. In industrial technological culture, the main force of valorization comes from capitalism, which turns significance into an expression of exchange value. However, one could argue that today, sign value has taken over from exchange value (Baudrillard, 1993). Valorization in terms of sign value wields an enormous 'presence' in virtual environments and symbolic markets.

En-presenting without valorization is impossible. This can be seen as the problem of waste. Recycling only became widespread once it was made clear that even waste has a value, even if it is negative, i.e. when the attribution of value concerns the removal of waste.[4] The point is not that there is no presence without value, but that absence is marked by a lack of value. What has been concealed by modern technology, for example the technological enframing of structurally induced needs, is not considered to be of value. It is made absent by a sheer lack of valorization. This is not the same as negative valorization, because even negative valorization (e.g. of waste) is a valorization (of its elimination). Lack of valorization is a far more subtle and insidious force of exploitation because it seems as if nothing is being exploited. Air, water and soil were all once seen as 'void' when they were heavily exploited as resources. It is only because of pollution that we have begun to valorize them, which entails a turning to facing 'the limits of growth'. Now they come with a price (perhaps still much too low), and this has set limits on their exploitation.

Indeed, it is not enough only *to know* what matters. For risk perceptions to obtain closure (and become risk assessments), they must entail an attribution of value (cf. Kraft, 1988). This is the transformation of utterance into understanding. Understanding must not be understood primarily as a cognitive concept, but in terms of a much broader hermeneutic of sense-making which includes a sense of *how much?* This thus also entails a notion of 'propping up' or 'supporting'. Without the attribution of value, we would not be able to deal with the practical problem of information overload, nor would we be able to direct appropriate resources to deal with risk perceptions. Closure requires a mode of selection that is 'proper' to the prevailing (inter)systemic logic.

Technoscience is nothing without a continuous input of resources from business and commerce, and legitimation from government and general public (Webster, 1991). In order to secure the continued mobilization of these valued types of capital, it must submit itself to the various processes of capital transformation. As a result, the products of technoscientific research can be expressed through the various economic, political and cultural exchange flows (Basalla, 1988; Rose, 2000).

Latour and Woolgar (1979: 192), for example, describe how the laboratory is organized on the basis of 'cycles of credit'. This enables them to draw connection between the rather instrumental rationalizations by scientists of their own motivations (in which credit refers to both financial reward and acknowledgement (respect) from peers which could be exchanged, shared, stolen, accumulated or wasted), with a vastly more diversified set of practices in which credibility operates outside the logic of monetary exchange. 'Credibility... concerns scientists' abilities actually to do science' (Latour and Woolgar, 1979: 198). Consequently, valorization (here seen as equivalent to the process of generating credibility) has to be seen in much broader terms than the pursuit of individual interests, it also entails power, belief and importantly, fidelity. It enables one to think of the materiality of networks – the flows through which they become 'established' and (to some degree, but never completely) 'fixed'. 'The notion of credibility makes possible the conversion between money, data, prestige, credentials, problem areas, argument, papers, and so on' (Latour and Woolgar, 1979: 200). Hence, it is not credit or recognition as such that motivates scientists but 'the acceleration and expansion of the reproductive cycle which produces new and credible information; that is, information for which the costs of raising an objection are as high as possible' (Knorr-Cetina, 1981: 71).

Risk production and risk management constitute an ensemble of technoscientific practices that require a matrix of visualization, signification and valorization. Through this matrix our being in the world is granted 'insight', 'meaning' and 'value' and becomes 'insightful', 'meaningful' and 'valuable'. Hence, rather than a fashion or a trend in late-modernity, technological culture has always been deeply imbricated in modernity; it has provided the sense and sensibility of modern thought from its inception in Cartesian metaphysics and Kantian ideology critique.

We have seen that technological culture offers particular types of social and symbolic organization of risk perceptions and risk sensibilities. However, it does not exhaust the ways in which risks are conceptually and practically engaged (performed). It is the excessive nature of risk – risks are forms of becoming – that prevents them from being totally framed within technological culture; they are a remainder of industrialized technoscience and a reminder that our attempts to programme and determine our own future, through the rational deployment of science and technology, always entail the possibility of undesired effects that remain unaccounted for. Hence, risk is that which – at least to some extent – remains beyond insight, meaning and value.

This is effectively the kernel of modernity. It may be seen as the socio-cultural manifestation of a drive to exclude and overcome contingency. However, since no rule includes the rule of its own application, the techno-logical predicament of such a drive will always remain paradoxical: the more contingencies are being incorporated, the more contingencies will be unleashed. Indeed, the more technological culture aims to control risk, the more risks will spin 'out of control'. The management of risk requires knowl-edge, but more knowledge also produces more risks. Hence, risks may be seen as a virulent force within modernity; they replicate themselves inside the technocultural bodies of modern society, by using the very same material that these bodies consist of. The 'other modernity' of the risk society that Beck was searching for becomes exposed as highly pathological, virulent and self-destructive.

We have demonstrated that technology reveals and orders; it is engaged with a form of en-presenting (making present, creating a presence) that consists of three modalities: visualization, signification and valorization. We have also argued that modern western technological culture is dominated by a specific form of en-presenting: representation. On the basis of the aforementioned analysis, one could suggest that only when technologies of visualization, sig-nification and valorization can be made to work in some form of assemblage (see Chapter 4) will it make any sense to speak of representation. Without some coordination (which does not mean an absence of ambivalence) between the attributions of sense, meaning and value to a virtual object, it is impossible to determine whether one refers to the same virtual object. That is to say, whereas we may assume that there will be multiplicity, there must always be a disclosure of means of transcoding from sense to meaning to value and vice versa. If any one of these three elements is missing, no virtual object can be actualized, nothing will be 'made present'. The different techniques that produce different manifestations can only be brought to render an account of the same virtual object 'it', if the ordering of 'it' can be decoded and encoded in a new language by a singular machine (an example of this would be the role of digitalization in telematics).

Finally, we have used Heidegger's discussion of technology as the conceal-ment of unconcealment to argue that Beck's concept of the risk society implies a turning. Whereas Heidegger was mainly writing about the effectiveness of modern technology and alerted us to the smooth operations of our technologi-cal culture, Beck teaches us something about the limits of these smooth opera-tors. Using Lodziak's notion of 'structurally imposed needs', we have argued that technological culture engenders a taken for granted of the ordering and reveal-ing of technology that turns itself into a dependence on such orderings. This dependence becomes manifest only when things break down, that is, when technology suddenly and unexpectedly loses its fidelity.

The Four Riders of the Apocalypse

To illustrate what may happen when technological culture breaks down, Part II of this book will focus on four different examples. The examples relate to the Four Riders of the Apocalypse as described in the Bible by St John the Evangelist in the Book of Revelation (6: 1–8): natural disasters (such as famine, floods, earthquakes and droughts), pestilence, civil unrest and war, albeit with some modification. 'Natural disasters' will be translated into 'waste', as it is becoming increasingly clear the current global ecological crisis is a crisis of excess. Pestilence stays the same, which is itself remarkable in the face of advances of medical technoscience. Crime will be narrowed down to 'compu-tercrime', or more specifically cyberrisks, which increasingly become one and the same with the virulent forms it relies on for its effectivity. Finally warfare becomes 'collective violence' which – because of the development of the modern state apparatus – no longer relies on a sharp dichotomy between order and chaos, but constantly iterates between the two.

The main purpose of these examples is not to prove that the end is nigh, but to actualize the potential significance of theoretical conceptualization in under-standing 'real' social and political issues related to risks in different techno-cultural constellations. That is to say, they are not to test the theories or prophecies, nor to directly apply them to 'reality'; instead they are to reveal modalities of thought in more practical and everyday descriptions of (relatively) 'ordinary' risk phenomena.

The notion of risk in Beck's work predominantly evolves around potential large-scale ecological catastrophes. His index case is nuclear risks, but the same arguments have been applied to genetics (Beck-Gernsheim, 1995) as well as biochemical risks (e.g. Adam *et al.*, 2000). These risks are similar in nature. They all stem from 'Big Science' projects that have emerged out of develop-ments in twentieth-century technoscience. These projects are marked by enor-mous investments from state and private capital, often in cooperative ventures. They emerged from new developments in science that were hailed as major breakthroughs, and potential epoch-shifting forces of progress and emancipa-tion. Furthermore, the risks associated with these large-scale projects are incal-culably large, and the consequences impossible to map and predict. In short, such ecological risks are dramatic, exceptional and spectacular. They signify hazardous ontological predicaments, not simply of specific groups of people, not even of the human species as such, but of life as a whole.

However, if we are to make sense of risks in a technological culture, especially when incorporating perspectives other than the macro-sociological risk society thesis, such dramatic catastrophic risks perhaps do not give the best empirical substantiation of our theoretical reflections (Scott, 2000). That is to say, if we are to understand the implications of risks, a focus on ordinary everyday life situations is more adequate. On this level, ecological risks are perhaps not the most ubiquitous, social risks such as traffic accidents and crime are more prevalent. However, many ecological risks and biorisks do enter the everyday lives of many people, for example when related to chronic illnesses and infectious diseases (McMurray and Smith, 2001). Indeed, many biorisks have an everyday presence and leave deep imprints on the social constitution of everyday lives, especially of those most immediately exposed to perceived risk conditions such as inadequate housing, poor diets and lack of hygiene. In contrast to the 'universal risks' of nuclear and genetic catastrophes or global terrorism (what Virilio (1986) named 'generalized accidents') which affect us all, these risks are far more socially selective and primarily target those at the bottom end of the socio-economic scale: the poor, the homeless, refugees, in short, the socially excluded, among whom many women, non-White groups and ethnic minorities.

Even if class still matters in the risk society, Beck is right in arguing that risks do not stop at class boundaries. The same applies to gender and ethnicity which – though clearly marked by asymmetrical risk relations with differential risk accumulations between men and women, Whites and non-Whites, etc. – are never simply defined by them in a singular dimension of have versus have-not. This is because the accumulation of 'bads' in asymmetrical risk relations does not always follow the logic of domination through which these bads are engendered. Whereas vested interests, especially access to resources, can be effectively deployed in the manipulation of both exposure to and circumvention of risks, they do not cover all hazards, as these also include unknown or not-yet identified bads. Moreover, being 'on top' in terms of social power may itself generate risks that specifically belong to such an exclusive domain of social relations.

As we shall see in all the chapters of this second part, risks have an innate tendency to transgress social divisions. They may originate from certain specific class, gender or ethnic locations, and these originations may show up in social divisions of initial risk exposure, once set into work, risks continue to affect other social groups. A classic example would be tuberculosis, which is currently on the increase. Although classically a disease of the poor, as it is related to poor housing, improper ventilation and a lack of hygiene, tuberculosis also claims its victims among the ranks of the wealthy who, after all, cannot always avoid to breathe the same air as the poor. Urban areas in particular are domains in which diverse social classes meet and share their microbes (Craddock, 2000). Microbes do not inherently prefer poor people. In perhaps a simpler vein, although the poor are far more likely to be victims of violent crime and are less able (resourced) to protect themselves against it, the effect on the rich has been equally dramatic as, in many urban areas, they have confined themselves to secure units, patrolled by private security firms, dogs, closed-circuit television (CCTV) and – in some countries – even weapons (Davis, 1990). Indeed, the anticipation of social risks already defines the 'victims' of risk relations. When you are rich, you have more to lose. This may sound as a slave morality, but it adequately depicts the liberal angst of middle-class White men who fear the wrath of ressentiment (a feeling of inferiority combined with a desire for recognition and moral revenge). Biorisks and social risks are public concerns which do not always match the relations of domination marked by social class, gender or ethnicity, and the collective political responses to them reflect this.

In the following chapters, we shall therefore focus more closely on specific and perhaps less spectacular kinds of ecological and social risks – those related to waste, disease, computer crime and collective violence. We shall see that since the Second World War they have all experienced an ascendancy in terms of risk perception. Moreover, they have mobilized technoscientific, governmental, mass-mediated and, last but not least, commercial responses which have signified them in highly specific ways which – instead of eliminating risks – have merely exacerbated risk sensibilities.[5]

Part II

The Four Riders of the Apocalypse

6 Cultivating waste

Excessive risks in an economy of opportunities

In this chapter we shall deploy the triad of risk society thesis, Actor Network Theory and biophilosophy to theorize waste. The purpose of theorizing waste is to show how what seems to be merely a 'side-effect' of the capitalist political economy not only poses serious problems for the sustainability of life in a capitalist world order, but – in a paradoxical moment – also transforms the very nature of that order into what Beck (1996) referred to as 'the world risk society'. As waste becomes big business, in terms of both financial and symbolic capital, the logic of 'goods' and its associated disciplinary regime become subverted. This creates an ambivalence in which waste itself accumulates, both in terms of valorization and pathogenicity.

The chapter starts with a brief assessment of the relationship between waste and risk, arguing that the inherently excessive nature of risk makes its 'objectification' (in science and engineering, as well as commerce and governance) an incredibly volatile process. We then look at how this volatility affects the way in which waste has been politicized in the age of industrial modernity, before focusing on the 'turning' towards what shall be referred to as 'the biopolitics of waste'. This 'turning' has affected the way in which technological culture has responded to the risk challenges revealed by technoscience, and has specifically evolved around the commodification of waste. That waste is not just big business is subsequently illustrated by drawing on Nigel Clark's work on 'nanotechnology'. Using his critique of some interpretations of the risk society thesis, the chapter concludes with a reconceptualization of 'waste-risk' as a virulent abject of modernity, poised to inaugurate a pathogenicity that is at once ephemeral yet deeply disturbing.

Waste and risk

Waste is perhaps the most universal, vulgar and banal example of ecological risks in everyday life. Nearly all ecological risks relate in one way or another to waste, and more specifically pollution. Whether nuclear, genetic or biochemical, ecological risks are materially engendered by pollutant waste objects – atoms, genes, molecules and cells. However, on a more mundane level, waste is less differentiated (even if separated) and appears as one amorphous blob of matter out of place (Douglas, 1966/1984).

Waste is often linked to disease (W. Marx, 1971; Wills, 1996). It represents uncontrolled matter-out-of-place, freely interacting and reacting, cultivating bacteria, fungi and toxins that may pose direct threats to our health. Ever since the connections had been made between germ theory and the sanitation movements in western Europe and the USA in the late nineteenth century (Brüggemeier, 1989; Latour, 1988), the regulation and expulsion of waste from the domestic and public spheres has been a governmental priority, although not everywhere as Craddock's (2000) historical account of San Francisco's rather lax regime of public health management illustrates. In many ways even today, public health management still espouses the values of the nineteenth-century sanitarians. The introduction of closed sewage systems exemplifies the articulated need to separate waste flows from all other flows in the domestic and public domains (Roderick, 1997). The removal of waste-dumps from urban areas, and from areas of human inhabitation generally, equally signifies this need to separate waste from ordinary life (Cope *et al.*, 1983; Cutter, 1993; Hofmeister and Spitzner, 1999). Beck's risk society thesis, discussed in Chapter 2, strongly evolves around associations between industrial modernization, waste and tangible and intangible risks. Waste, produced as side-effects of industrial production, engenders latent risks to human health.

A clear example of this is the relationship between waste and disease. First there is a direct link between waste and infectious disease. Typhoid fever, cholera and gastroenteritis are all caused by types of bacteria that live on (human) waste. When human waste is not removed from drinking water, such infectious and virulent bacteria can easily reproduce themselves. Cultivated in the sludge of human and animal excreta, water and offal, they pose an immediate threat to human health. Needless to say, their presence is most ubiquitous in 'open' sewage systems (Wills, 1996). Second, the relationship between waste and disease can be indirect. Here we should particularly focus on the role of animals and parasites such as rats, mice, fleas and insects. *Yersinia pestis*, or the Plague, for example, is carried by rat fleas. Rats and humans often cohabit, especially when there is a lot of waste around. Cockroaches and houseflies are equally prolific carriers of diseases and they too thrive on waste.

It was against such kinds of risk associations with waste that the earliest public health reforms were initiated in the nineteenth century. These initiatives, spearheaded by health reformers, hygiene movements and medical scientists, were the primary forces behind the widespread introduction of sewage systems, organized refuse collection, incineration and the use of landfill dumps outside areas of human habitation.

However, not all waste-risks are that easily overcome. Especially risks related to the toxicity of certain chemicals of matter-out-of-place have proven to be more difficult to remove (Cutter, 1993). Sometimes, however, these risks did become spectacularly manifest – as in the Chernobyl (1986), Bhopal (1984), the *Exxon Valdez* (1989) and *Torrey Canyon* (1967) disasters. In all four cases, the relationship between industrial production, waste and risk were revealed in the dramatic consequences on the environment. Furthermore, the fact that in these cases waste-risks were inseparable from the nature of 'goods' produced by the industrial production itself, served as a reminder that waste-risk is never simply an undesired side-effect of modernization, but endemic to it (Allan *et al.*, 2000a: 8).

More often than not, however, the endemic nature of risks hardly manifests itself. In this sense it is a typical example of Beck's argument that risks thrive in obscurity and denial. Waste, defined as 'matter-out-of-place' or – more normatively – that which is to be discarded, is often placed out of sight, made to disappear, often by simple means of removal (refuse collection, landfill dumps, incineration). But many types of waste are intangible. Nuclear radiation, for example, escapes all human senses. Operating beyond the senses of common perception, risks vanish and become reduced to hypothetical speculations.

The principle of out of sight, out of mind has for a long time been useful in keeping the lid on the negative side-effects of industrialization. During this period, toxicity was allowed to build up in the soil, in the air and in the water. Only sporadically would its side-effects be noticed in terms of epidemiological anomalies of, for example, clusters of chronic illnesses, cancers or miscarriages (Colton and Skinner, 1996). Unlike the spectacular examples of accidents and catastrophes, these largely escaped the news media. For example, as early as the 1930s, medical scientists had been concerned with health risks due to prolonged low-level exposures to toxic chemical substances (W. Marx, 1971: 38) and radiation (Caufield, 1989; Ritenour, 1984). The problem of low-level exposure is often exacerbated by the lengthy timescales of cause and effect, and by the latter's often dispersed and fragmented manifestation. Yet these same conditions made scientifically based risk assessments virtually impossible because cause and effect do not operate on the same linear timescale. The effects appear too dispersed, too contingent, too haphazard, to be all tracked back to a single cause (Adam, 1998).

Refuse waste is now generally considered to be a social problem. Our society produces immense quantities of waste. All waste needs to 'go' somewhere. That is to say, the tried and tested strategy of refuse, based on the principle of out of sight, out of mind, so well represented by the bin, no longer works. Waste – as matter out of place – still 'takes place'. It does not suddenly disappear once it has been discarded. As anyone living near a waste-disposal site knows, waste can be seen, it smells and sometimes it spills over in the form of scandals, such as that of Love Canal (Colton and Skinner, 1996). In this scenario, waste is still 'that which is to be discarded' but it has now become 'that which can never be completely disposed of'. Waste somehow returns as a problem, an irritating nuisance, a health hazard or an ecological disaster.

When issues over toxic waste in water basins and landfills started to surface in the 1970s, a floodgate was opened in terms of environmental policy issues. Suddenly governments were faced with problems that were the consequence of actions taken years ago, usually within the limits of law. The companies who

had legally dumped toxic waste and were responsible for the negative consequences, however unintended, were thus legally unaccountable for their
actions. Those who suffered the consequences experienced the reverse: full
accountability without responsibility. Decades of economic progress were paid
back through prolonged human suffering.

Waste exposes the fundamental problem of time in modern society (Adam,
1998). It has re-entered the product life cycle as 'matter' from a completely
different time zone. That is, waste is being en-presented – almost in a zombie-
like fashion – as a past returning to haunt the brains of the living. This en-
presenting of waste first manifested itself in terms of a scarcity of 'waste-space'.
As landfills at the edge of urban enclaves were being enclosed by the suburban
sprawl of the 1960s and 1970s, the chemical toxins – thought to be 'gone' –
were seen no longer as local epidemiological anomalies but as nation-wide
statistical aggregates. That is, once they were being 'embodied' by human beings
and attached to the immutable mobiles of statistical epidemiology, waste no
longer disappeared into complete absence, but became a reminder of hazards
and a herald of a new age of risk.

By its very nature, waste inhabits the ambiguous borders of 'presence'. As that
which is out of place it calls on a need for its removal. By definition, waste
equals excess. It is excess production, excess material, leftover, dysfunctional,
the product of inefficiency. It has an elusive, even ephemeral nature. Yet as risk
it manifests itself ineradicably in the public and private domains of modern life.
As excess, waste thus engenders a volatility. Hovering between presence and
absence, its ubiquitous ephemerality poses serious problems even for a culture
based on strong boundaries of inclusion and exclusion. The excessive nature of
waste returns to haunt the present as past-waste becomes present-toxicity.
Suddenly almost all ecological risks could be rearticulated as pollution or
waste-risks; indeed, one could argue that the world risk society is above all a
world waste society.

Politicizing waste

Despite its apparent dysfunctional and excessive nature, waste has operated in
close symbiosis with human social organization long before the advent of modernity. Mary Douglas's (1966/1984) famous work on the association between
dirt and taboo gives an excellent account of how waste operates as *nomos* – a
principle of regulating flows through distribution and boundary setting – for
both human bodies and the abstracted social bodies across different types of
society, and is thus a site of 'nesting' for the incorporation and incarnation of a
more complex social order. As Douglas has shown, waste always refers to *more*
than the matter that is out of place. It is a matter of boundary setting, maintenance and transgression.

In the early modern conceptions of 'proper' social order, its out-of-placeness
took on additional social-symbolic value of signifying moral deficiencies
(McCarthy, 1997). The associations between waste and disease resulted in

often quite radical innovations in urban planning and public health regulation. Waste-risks were often the catalyst of emergent alliances between medical science, public health management and national economic and political social interests. Governmental regimes of disciplinary power were formed in close proximity to technoscience (Donzelot, 1977). For example, Ali de Regt's (1984) study of the emergent disciplinary regimes of social workers in the Netherlands during the nineteenth century illustrates how morality was redefined in terms of properness, which was bound up with a rationalization of the use of space, time and resources. The reduction and removal of waste (including 'wasting time' and alcohol use) was given top priority as the middle class sought to expand their hygienic regimes onto the working class and underclass. As hygiene became a central project of nation-building and the reinvigoration of a healthy and disciplined national population (especially in northern Europe), waste became politicized as antisocial, i.e. as an enemy of the people (see Latour, 1988).

We modern human beings have learned to live with fairly complex systems of waste-flow management. Starting with the closed sewage systems, waste-flows have been increasingly regulated, channelled and controlled by both incentives and punitive measures since the advent of modern, disciplinary society in the nineteenth century. A main driving force has been risk aversion. In the domestic sphere, the main place to become a site of risk-aversion strategies was the kitchen (McCarthy, 1997). The kitchen is an unusual space because it is the place where substances are prepared that are designed to enter the body (food), and thus a potentially dangerous zone where 'bads' can enter the same body as well. The kitchen consists of a continuous dual flow of 'desirable' and 'undesirable' substances, which often stem from the same source. That is, waste becomes waste only once it is being removed from the original entity, when the 'good' has been separated from it. The splitting of a product into good and bad engenders the image of waste as an undesired side-effect. However, this obscures the fact that what are called goods are easily and rapidly turned into bads. As we all know, many foods decompose quickly and become themselves sources of bad smells and breeding grounds of bacteria and parasites. Food/waste presents an excellent example of the volatility of risk. This volatility has profound implications for the way in which we manage risks.

It is therefore a gross mistake to think of the problem of waste as a recent phenomenon. The pathology of waste has been endemic to modernity, and has never been exclusively relegated to complete 'absence'. However, what has been a recent phenomenon is an acknowledgement that waste is more than a mere negative side-effect of modernization. Whereas hitherto all forms of industrial modernity have thrived on excessive waste-production and re-production, it was only since the late 1960s that waste itself became articulated as a global ecological problem (W. Marx, 1971). This sensibility was strongly and publicly reinforced with the debate around the *Limits of Growth*, initiated by the Club of Rome in 1972 (Meadows *et al.*, 1972), that explicit links were being made between industrial modernization and waste production, at least in official governmental discourses (Macnaghten and Urry, 1998: 46). This took place alongside a rapidly growing field of 'environmental science', which was highly concerned with issues of biochemical pollution and their impact on the natural environment and human health (e.g. Farvar and Milton, 1973; Mellanby, 1972; W. Marx, 1971). Most central here were concerns about pollution of air, water and soil and particularly the smog of urban centres such as London and Los Angeles became famous visible markers of the ineffable and growing problem of waste.

However, whereas the sciences were producing vast amounts of evidence about environmental risks related to waste, they were unable to generate an effective political and governmental response. Low-level exposure risks are notoriously difficult to prove and apart from the legal minefield of attributing liability through the deployment of scientific expertise, science itself was far from unequivocal about the relationship between wastes and risks (e.g. Mellanby, 1972; Yearley, 1991). Public concern was similarly haphazard, often related to the rhythms of media concern and moral panics. The general official stance on waste that as long as it was out of sight and separated from the 'normal flows' of domestic and public life, we need not bother with it too much, was able to continue relatively undisturbed despite the early warnings from emergent environmental movements (Slee-Smith, 1976).

However, especially over the last decades, things have been changing rapidly. Since the 1980s, the problem of waste has been more consistently addressed on several political agendas, including those of the right, who initially found it more difficult to attune to ecological risks as political opportunities (Macnaghten and Urry, 1998; Yearley, 1991). Perhaps it is because (household) waste is more easily articulated as a moral issue than other ecological issues (e.g. resource depletion or industrial emissions), that the political right was able to incorporate it into its own discursive formations without upsetting the main strategic core (e.g. supporting capitalism with nationalist moral authoritarianism). This remainder of the nineteenth-century sanitarian movement, who in those days were very much part of a liberal, emancipatory and progressive social movement, has been incorporated smoothly into environmental political strategies that do not automatically imply a criticism of industrial capitalism. What comes under the banner of conservationism has enabled an articulation of waste-risk as both an individual and social issue. Conservation, or not-threatening the status quo, implies attempts to preserve the idea of a natural balance (Dobson, 1995). Whereas some waste could be seen as part of this balance (e.g. 'muck' as fertilizer), the general idea is that there is now too much waste and something needs to be done about it.

An interesting aspect of this conservationist line of environmental politics is the collocation of waste and litter (after all, litter is matter-out-of-place). Waste-as-litter signifies undisciplined carelessness with what has been left over, a disregard for one's environment including other people. As a society of affluence has a lot of leftovers, due care would be needed with these if the affluence itself cannot be addressed. Litter is a moral issue; it signifies lack of concern, lack of discipline, will-power, etc., which incidentally parallels similar symbolism with obesity (Bordo, 1993).

Hence, with virtually no opposition, waste became a primary target of environmental risk politics, irrespective of the political colour of the government (needless to say, this colour still mattered a great deal with regard to the kinds of strategies of waste reduction that were being developed). Indeed, anti-waste politics – e.g. the depletion of the ozone layer due to wasteful emissions of chlorofluorocarbons (CFCs) – became virtually hegemonic (Macnaghten and Ury, 1998; Yearley, 1991).

The politicization of risk, which has been an ongoing process since the nineteenth century, thus underwent a new turning. Associations with 'environmental issues' and *thus* with waste, became viable marketing strategies for both electoral politics and commercial enterprise. As with early public health reforms, this hegemonic turning is predominantly made within technological culture. That is, it is directed at strengthening the existing systems of visualization, signification and valorization. This is not to say that everything runs smoothly. Far from it. But the modes of resistance, especially those articulated with the excess of waste, have not been central to the emergent strong actor networks at the core of politics of waste.

Technological culture's response: a biopolitics of waste

For more than two centuries, modern 'man' sought to control nature, divinity and history through technoscience, which evolves around 'rendering an account' of the laws of nature. These laws may be invisible, but can be made to appear through the spectacle of 'demonstration' (Latour, 1988). In doing so, however, technoscience unwillingly and slowly sows the seeds of suspicion and the collapse of its own claims to a sovereignty of truth. It has become its own hostage to fortune. By revealing what is not-yet visible, it holds out the promise that there is always something more to be revealed. It thus cultivates a sense of the excessive nature of truth. That what has not yet been rendered visible thereby returns to haunt the very magic of technoscientific disclosure. Through that demonic beast of suspicion, the excessive nature of truth turns from benevolent spirit to an angel of obstruction.

Technoscience, on an infinite quest to reveal what has remained hidden, opens itself up to suspicion and doubt about what it does not yet know. It is therefore that in its fragmented and dispersed practices of 'trials', it has abandoned the grand narratives with which it set out to legitimize this quest. The meta-narratives have been abandoned a long time ago, well before their incredulity was acknowledged under postmodernism. Instead, it was a loose association of 'trials', lumped together by an abstract signifying system called 'positivism' which – instead of describing science – was engaged in Science as a politics of networking, inaugurating cycles of credit and rendering specific resistances obsolete while making others transcendent. What actually goes on in scientific practices rarely corresponds with its idealizations under the label of positivism (Latour, 1993).

Whilst being at the heart of the Enlightenment, positivism was also instrumental in the destruction of Enlightenment ideals. It harnessed an instrumental and fundamentally irrational reasoning that systematic observations could induce a reality, usually in the form of regularities that could stand in for (represent) laws of causation. However, despite its logical impossibility, this ethos – which is not at all scientific and thus not *practised* as science – still finds its practical applications in a technological culture governed by pragmatism and a taking-for-granted of common sense. Although positivism may be dead as a philosophy of science, it still rules supreme as Science (Gilbert and Mulkay, 1984; Yearley, 1991) including scientific approaches to risk (Adams, 1995; Cutter, 1993; Lupton, 1999; Wynne, 1996).

This residual positivism embraces a pragmatism and valorization of common sense, without the epistemological pretensions of its nineteenth-century predecessor. It is really a manifestation of performativity (Lyotard, 1979). Performativity does not need justification or legitimation. It justifies itself in action by producing effects and maximizing efficiency. It affirms itself by assertion. It is above all a pragmatic response to a belated revelation that it never was what it claimed to be: an apology for what already exists. Hence, such pragmatism cannot pose a radical alternative to the waste-risks of the ecological crises it faces. This becomes clear in the concept of sustainable development as it seeks to ground itself on a scientific justification of acceptable levels of wasting and is thus better described as a 'management of decline' or 'damage limitation'. This is because with the abandonment of grand narratives, its performativity merely generates a rhetoric of values that is ultimately arbitrary and can be changed at will. There are no fundamental values against which it has to be tested.

However, from the sociology of scientific knowledge (SSK) we have learned that scientists continuously make evaluation and deploy value-judgements. As Yearley (1991: 124) has convincingly shown, '[scientific] judgements inevitably go beyond the factual evidence on which they are based and . . . facts themselves do not enjoy an unassailable status'. He continues by describing the uneasy alliances between science and environmentalism to argue that the latter's increasing reliance on the former has made it more rather than less vulnerable to dismissal. Because of the moral indifference of the scientific method, Yearley argues that it is unable to ground its claims on any other basis than 'facts'. Especially when these facts are intangible and immersed in highly complex issues such as ecosystems, uncertainty and undecidability prevail. The lack of scientific closure can easily be exploited by judicial, economic and political powers to dismiss the environmentalist's claims and concerns.

There is, however, a serious flaw in this reasoning. Yearley's empirical evidence of the limitations of scientific closure due to lack of evidence is unduly constrained by his own idealist conceptions of science. In the tradition of SSK he sees science primarily in terms of rhetoric: arguments, opinions, statements and discourse. The only trial he is willing to consider is that of scientific authorization itself. Confusing Science as epistemology (positivism) and science as practice, he thereby glosses over the relatively simple observation that science essentially evolves around, yes even thrives on, uncertainty. The absence of facticity, the lack of closure, is the opening up of cycles of credit. By insisting on a rather narrowly defined understanding of 'authority' as informed by the ability to make a legitimate claim to speak about something, he fails to come to terms with the inherently political nature of any act of signification and representation. It is not automatically the case that, when there is insufficient evidence, scientists will not speak; often (one could even say by definition) scientists do speak prematurely. The question is to understand why and under what conditions scientists invoke uncertainty or heterology to defer judgement.

Most importantly, Yearley fails to acknowledge the active nature of 'facticity' or 'objectification'. 'The' environment is never a 'thing' to be operated upon, but always 'activated' and thus itself 'active'. Contingency – the active role of objects and contexts – is not an enemy of scientific authority but a powerful ally (see in the case of nuclear waste, for example, Caufield, 1989; de la Bruhèze, 1992; Ritenour, 1984; Welsh, 2001). This is because authority is not exercised in the fullness of knowledge but, instead, by the lack of it. This lack can be both due to absence or excess. In the next chapter we will argue that such authoritative closures are increasingly invoked by an excess of knowledge.

Rather than a lack of closure, then, one could argue that environmental politics suffer more from too much closure. By becoming hegemonic, especially when cluttered around that ambivalent slogan of 'sustainable development', many claims are being lumped together. In the previous chapter, we have stressed that risks have engendered a turning in technological culture. This has not been an immediate revolution, but a rather slow one that only recently picked up speed. Whereas the argument thus far focused mainly on the crucial role of temporality and anticipation of risks as being-towards-death, we can now add another, more mundane-political dimension to it. What was significant about these new risks emerging in the 1980s just when capitalism's grim face started to exercise an almost arrogant hegemonic control over the world, was that for the first time since the Industrial and French Revolutions a rupture occurred in the association between survival and (self-centred) individualism.

Throughout modernity, the predominant concept of 'modern man' has been that of the *homo economicus* of liberalism. As a first principle of utilitarian rationality, a society in which everyone acts as the homo economicus would always find a collectively optimal arrangement through the aggregation of individual, self-interested acts. Hence liberal economics, for example, has always argued that collectively based arrangements are at best temporary interventions, a necessary evil only required until the market would finally take full control. Thatcherism in the UK or Reaganomics in the USA could be seen as culminations of this philosophy, promoting harsh individualism, justifying greed, dismantling collective arrangements, simply to eliminate obstacles for the full realization of utilitarian rationality.

The essence of Beck's risk society thesis is that such reasoning simply does not apply to risks such as nuclear and genetic technology or pathogen virulence. Here utilitarianism fails because the individual simply cannot obtain security by following the logic of individual self-preservation. Survivalism is powerless against the kinds of catastrophes predicated upon us by these aforementioned risks. No market can ever be expected to effectively deal with nuclear radiation, genetically modified life or pathogen virulence.

Whereas for almost all other political ideologies (including communism and fascism), and most religions, this is nothing new and can be extended to many other social problems (crime, poverty, 'moral degeneration'). Thus risks often have the most devastating effect on liberal pluralism because they immanently undermine the logic of the calculating, self-interested, utilitarian. It turns the individualist ethos of self-preservation into the same camp as the forces of death. Utilitarian rationality becomes suicidal.

This is most remarkable. After all, what could be a stronger assertion of life than survivalism? The association between survivalism and life, however, only works under conditions of scarcity, where greed could give one (an individual, tribe, pack or herd) an edge over one's neighbour-competitor. As soon as the 'bads' exceed the goods, i.e. the problem is an abundance of bads that cannot be compensated by goods, greed becomes self-destructive. Again, this phenomenon is in itself not new. A history of epidemics testifies to this (e.g. Horrox, 1994; Carmichael, 1997), as does the threat of total nuclear wipe-out during the Cold War (Allan, 2000). What is new, however, is the juxtapositioning of a range of such risks together with the apparent inability of utilitarianism to sublate these through the concept of self-centred rational choice (of markets and electoral democracy). Instead, what was called for was a return to more state-led social organization, governance and socialization, with commerce becoming more strongly regulated, especially in terms of risk production.

New alliances between technoscience, governmentality and commerce had to be formed. Sustainable development proved to be a performative and affirmative virtual object across all three domains. Waste-risks played a crucial role in the formation of this emergent network. It became the enemy against which it could be unified. Waste-risk was the immutable mobile that facilitated new cycles of (scientific, political and commercial) credit. This new alliance is often represented in terms of 'the Third Way' – a merger between governance and commerce, strongly supported by technoscientific-based forms of reason (as well as drive to innovation). What is remarkable here is that the affirmative aspect of 'sustainable development' is a politics of life – a biopolitics. It poses a claim to necessity and urgency on behalf of the survival of the human species globally. Its appeal is thus all-inclusive (and is thus at odds with those of national interests) and hegemonic because it seeks to stall the alleged incumbent self-destruction of 'modern man'. It is a biopolitics that also affirms the collective and social basis of human life, and the mutual dependencies between human beings are extended by the species' overall dependency on the global ecology.

According to Beck, the risk society is primarily geared towards the institutionalization and commodification of particular kinds of risk, that is, their public domestication and transformation into regularities on the one hand, and their valorization and transformation into economic and political opportunities on the other hand. In this sense, the management of waste-risks is an allegory of the transformation of the public sphere, described by Habermas (1989) as being undermined by both instrumental rationalities of technocracy (and bureaucracy) and commodification. However, in this world, waste-risk becomes more than an allegory of the decline of democracy. It signifies the more fundamental problem of how to manage flows between seemingly autopoietic systems of production and consumption. If waste occupies the rather ambivalent position between matter and non-matter, if its materiality defies the metaphysics of presence, how can it be positively incorporated, regulated and controlled?

At the same time, waste today is big business. Refuse collection and waste management such as refuse collection is a major activity of (local) governments, involving a considerable share of the gross national product (GNP) of modern industrial societies. Recycling too is a multimillion dollar global industry, involving not only the richest countries in the world but, significantly, also the poorest (Yearley, 1996). Waste is indeed an essential part of the global capitalist economy. Its flows mirror those of capital although usually in the opposite direction.

As a response to the mounting costs of waste, the principle of sustainability is being built into the design and development of an ever-increasing number of products. Wasting is becoming more and more expensive, and the reappropriation of waste has become a cost-saving exercise as well as a good public relations marketing device. The so-called 'greening' of the economy, often embroiled with the rhetoric of sustainable development, has become a profitable political and commercial strategy of managing the decline of the world ecology. It does not challenge the logic of capitalism, the working of its markets, the relations of domination that underpin it, or the systems of governance that keep it in place; moreover it does not threaten the system of manipulating needs upon which consumer culture so heavily depends (Lodziak, 1995; also see Chapter 5). We can all continue, with technoscience ready to hand to give us more and more gadgets and devices for incorporating sustainability, i.e. by keeping 'waste' inside.

A clear example of the emergent waste-economy is the growth of the recycling industry (Slee-Smith, 1976). The technology of recycling may unconceal waste-as-matter, but conceals it again by enframing it as another standing reserve. For example, Holmes (1981: 37–43) advocates the use of waste incineration as economically viable, but spends only a few words on the waste (emissions, residues, energy used) produced by this process. Waste is thus constantly displaced (as that which is to be shed), but which always returns to haunt the holism of the present. It is still paid off either now or later.

The organization of waste management has always been a matter of regulating and securing waste. However, simple 'disposal' is no longer a viable option, as the negative consequences of that (including financial as well as health-related drawbacks) have already become unmistakably apparent. Instead, sustainable development entails strategies of ordering waste into new forms of functional applicability. In particular with recycling, the idea is that there is more use-value to waste than is currently exploited. The further exploitation of waste, it is argued, leads to a reduction of waste and the need for primary resources (Holmes, 1981; Managers Ontmoeting Overheid Bedrijfsleven, 1993; Van Loon and Sabelis, 1997; Welford and Gouldson, 1993). One could distil from this that if the technologies of recycling can be attuned to the technologies of production one could achieve sustainable development by damaging neither the capitalist economy nor its ecological infrastructure. Indeed, one could have one's cake and eat it (Dobson, 1995).

One particular domain in which this new optimism carries much weight is that of 'green business'. The Body Shop, for example, prides itself for its

> commitment to environmental and social excellence . . . The ultimate aim of The Body Shop is to include environmental issues in every area of its operations but, at the same time, the organisation rejects environmental opportunism which has often paralleled the green marketing strategies of more cynical firms.
>
> (Welford, 1994: 93)

As the Body Shop protocol clearly shows, business ethics and commercial success can be synchronized quite easily by marketing strategies that are specifically targeted to highlight the articulation between responsibility and choice. By offering green products, businesses can capitalize on a growing environmental awareness among consumers, and in the case of the Body Shop, the business itself has been quite active in helping to construct such awareness. The double loop of marketing as both inducing and responding to markets pays off well in a business ethos that thrives on selling a positively responsible lifestyle.

The environmental politics of green business (and green governance) requires a strong notion of accountability. The central role of polling (consumers, voters) and auditing activities that are an ongoing practice in the management of Body Shop testify that the organization favours forms of reflexivity based on statistical aggregates (Welford, 1994: 97–101). Whereas they are often justified either in terms of marketing rationale or democratization, the explosion of polls and audits are indicative of wider shifts in the relationship between commercial and political organizations and institutions and the public they claim to 'serve'. This shift towards 'trust in numbers' itself signals a changing modus of accountability and a 'decrease in trust in conventional notions of representation' (Macnaghten and Urry, 1998: 77), including those of traditional marketing techniques. In the case of the Body Shop, the continuous self-monitoring is claimed to be developed on the firm's own initiative as part of its identification with an 'ethical organisation committed to environmental improvement' (Welford, 1994: 96). Such self-monitoring is a pragmatist-positivist variation of reflexivity as self-confrontation. However, it is attuned to self-valorization rather than self-criticism.

The Body Shop is merely an index of a wider movement in ecological economics which, with the advent of sustainable development discourse, has steadily become a near hegemonic formation expressing the 'only path' towards an ecologically 'balanced' social order that is economically viable in late capitalist societies.

> The fundamental basis of this goal is a commitment to the broader concept of sustainable development which considers not only concepts associated with the environment, but also equity and futurity. It is the strong belief of The Body Shop that the moral burden of achieving sustainability in business should become the principal driving force behind business in the future.
>
> (Welford, 1994: 94)

From this it can be deduced that sustainable development involves the attunement of economic, juridico-political, socio-cultural, psychological and ecological 'systems' as a result of which all systems operate on a singular integrated functional plane as an expression of the intention and attempt to prevent all excessive damages to each (sub)system. The biopolitics of sustainable development are thus based on a principle of 'excess limitation', which is seen to be achievable via a combination of 'ethical commitment', 'rational planning' and 'continuous self-monitoring'. Sustainability is thus a philosophy and a cultural practice, a way of life that is integrated in science, politics, media and commerce.

However, that 'sustainable development' is a management of decline finds a most clear outlet in the management of waste.

> We believe that wealthy societies have an urgent and overwhelming moral obligation to avoid waste. As a responsible business, we will adopt a four-tier approach: first, reduce; next, re-use, then, recycle; and finally, as a last resort, we will dispose of waste using the safest and most responsible means available.
>
> (Body Shop's *Green Book*, cited in Welford, 1994: 95)

It is a moral imperative that waste is seen as something to be avoided, but – as even the Body Shop acknowledges – it will never completely vanish. Hence other operations are necessary. These are placed in an order of valuation: reusing is better than recycling and recycling is better than disposal. Responsible business is thus focused on producing less waste; but defining what counts as waste is not a matter of ethics alone. Science is called upon to differentiate between hazardous and non-hazardous waste. By presenting 'science' as an unequivocal touchstone of waste-identification, ethical business is able to justify itself with scientific legitimacy.

Whilst avoiding it, the biopolitics of sustainable development thus embrace waste as a resource for new recycling industries (Van Loon and Sabelis, 1997; cf. Holmes, 1981; Slee-Smith, 1976). Sustainable development presents itself as the only feasible alternative to the type of modernism that has brought us excess pollution and potentially catastrophic unforeseen risks and hazards. The ideas it generates allocate moral responsibility and decision-making powers to technocrats since the balancing of economic and ecological accounts is not a matter for politicians, let alone lay people. Ethics become subsequently subject to similar rational-synoptic planning mechanisms. Intertwined with technocracy, ethics remain little more than an empty shell, or worse a strategic smoke-screen to divert political pressure.

However, once appropriating waste 'as' something useful, this biopolitics of technological culture begins to encounter its own limitations. Sustainable development is based on the prevention of 'excess'; hence 'waste' as such is not a problem but 'excess waste'. However, following the same mould of representation, all waste is by definition excessive.

In sustainable development discourse, waste management is based on the projection of 'waste' as a present, useful entity that is ready to hand and be handled on a market, merely depending on our technological capacity to reprocess it into the chains of production. The problem with this is an inadequate conception of time. Waste is by definition a temporary phenomenon. Because waste is inherently excessive to what is made ready-to-hand, its transformation into functional matter still entails a process of dissembling and assembling, which takes time. The temporal framework in which the currently dominant logic of waste management is structured, however, implies a singularity of time as a technologically manipulable parameter. It does not allow any consideration of possible, unknown and undesired consequences and incubations that may happen in the meanwhile of the transformation from waste into a standing reserve.

No example is perhaps more compelling than that of nanotechnology. Nanotechnology is to a large degree still a hypothetical productive system whose aim is 'to arrange individual atoms into workable devices, machines which will then perform further operations on an atomic or larger scale' (Clark, 1998: 354).[1] Nigel Clark argues that a key motivator for nanotechnology is to intervene to renormalize pathologies in nature at a submolecular level. Its advances are that it can operate at immense speed and across a wide range of scales. Nanotechnologies are designed to become self-replicating to maximize speed and impact.

Especially in the face of mounting heaps of waste, nanotechnology offers an imaginative, proactive response to tackle the problem head-on. 'In the eyes of nanotechnologists, the environmental crisis is largely the result of the intolerably crude assembly techniques of the present industrial regime' (Clark, 1998: 358). Nanotechnology offers not only a vision of maximum efficiency in production, but also in the possibility of transforming waste back into functional matter. The 'loose molecules' or pollutants are to be reconnected. This is also the essence of biodegradability. By mimicking the 'natural' processes of molecular assemblages – a redefinition of life – nanotechnology offers itself as an integral part of ecosystems. Indeed, it promises an artefactual form of industrial modernity that is at the same time completely 'natural'.

Clark distinguishes between linear and non-linear hazards, with the first being largely confined to unintended consequences of particular human and artefactual practices and interventions and the second stemming from the complexity of ecosystems and information processing devices that span across an undefined time span and involve a variable number of actants. Nanotechnology promises to act not only on linear hazards (such as waste), but also on non-linear ones (such as emergent viruses). The specific problem with the second, however, is that the technological system must thus also attain a certain autopoiesis – of self-replication and cybernetic feedback loops. The unpredictability of non-linear hazards, as well as its speed and scale, has to be matched by nanotechnology. Thus, Clark argues, it is impossible for nanotechnology to be effective and remain under tight (human-centred) control.

Indeed, the horrifying aspect of nanotechnology is that it becomes indistinguishable from the superconductive events (ecological catastrophes) that it is called upon to combat. Like genetic engineering, nanotechnology thus entails the capacity of becoming itself a runaway event. As Clark shows, this is not simple a matter of problem-solving, but entails an 'aesthetic sensibility' similar to that of, for example, the artist Stelarc who wishes his work of art to become inscribed into the 'germ line' and thus attain a certain immortality (Caygill, 2000). This creationist ambition is an example par excellence of modern technological culture's assertion of the mastery of 'man' over nature and divinity. The hypermodern aesthetic of such radical creationism is diametrically opposed to the 'risk-avoidance' ethos propagated in the risk society, for example as sustainable development. It is an enthralment with risk and catastrophe that is also present in the work of some biophilosophers such as Nick Land (1995) who – taking an anti-humanist stance – celebrate and emphasize this as radical potential for the dawning of a new era. The question remains, however, whether nanotechnology is able to offer 'strange attractors' that pose a viable alternative to sustainable development. The more humanist-inspired risk society thesis may appear too slow, too much drenched in the toxic sludge of industrial modernity. But nanotechnology as an alternative form of biopolitics may equally – albeit more drastically – end in the destruction of all life forms. The biopolitics of nanotechnology may thus become a radical politics of death – a proliferation of pure waste.

Conclusion: waste as a virulent abject of modernity

Waste posits a problem for modern societies not only because it has become clear that our strategies to contain it have failed, but also because our conceptual grasp on it is slipping away from us. We moderns think and act upon categories that are based on dualisms such as general/particular, eternal/historical and temporal/non-temporal (Deleuze, 1968/1994). These categories constitute a metaphysics of presence. Waste defies all of those dualisms because it 'is' not. Its ontology is negative; an absent-present.

Waste reflects the pathologies of inefficient production – something that is itself already antithetical to the episteme of modern thought. Hence its current threat is more than a case of chickens coming home to roost; it offers new possibilities for an expansion of modernist megalomania, exemplified by some of the extreme cases of both 'panic ecology' (calling for ever-increasing and pervasive regulation) and anti-humanism (catastrophic enthralment) described by Clark (1997b, 1998). The shift from invisible absence to omnipresence (as a resource for recycling, ethical business marketing, political strategy, nanotechnological innovation, etc.), however, still inhibits an inadequate conceptualization of waste.

The repetition of waste, its eternal return, is not a matter of a fully enclosed circularity (reuse/recycling), but related to the modes of displacement and disguise that have marked (and marketed) its ontology from the beginning. It is the untimely, that element of time that can never be reduced to the temporal – neither via equivalence, nor via resemblance, nor via analogy nor via opposition – that marks the transgressive force of waste.[2] 'When the identity of things dissolves, being escapes to attain univocity, and begins to evolve around the different' (Deleuze 1968/1994: 67).

The problem of waste is a universal one. It is the problem of selecting what belongs and what does not, that is, it functions as a focal point for politics of inclusion and exclusion, with waste (as that which is left over and discarded) relating to the latter. However, waste always returns. What is being discarded is not suddenly no longer there, something remains, a residue of what once was a useful matter. The metaphysics of presence, which understands Being and Presence as identical, neither applies to risk or waste as their being is actualized only in absence. Hence, risk and waste are perhaps better understood as part of a metaphysics of absence. They remind us of the incompleteness of absence.

Whereas the return of waste in the form of risk has always been enframed in a metaphysics of absence (i.e. out of sight, out of mind); it is more recently that its return now also enters the metaphysics of presence. That is to say, whereas disposal used to be sufficient in realizing the wastefulness of waste, today even disposed waste remains present. This return of the same, or better, of the simulacrum of excess, has primarily been the work of risk. Hence, even disposed waste (a past event) now demands extra remedial treatment and management, primarily by means of governmental intervention (W. Marx, 1971; Slee-Smith, 1976) but also private initiative (Holmes, 1981). This treatment or management could be seen as attempts to reincorporate exclusions.

Whereas all forms of life exist in specific relationships with waste, in modern industrial western society, human beings have been faced with an immense amount of waste material. Indeed, waste has always been driving contemporary capitalism. Without active wasting, we would not 'need' constant intensification of product life cycles, as there was no need for absorbing ever greater amounts of surplus supply. Bataille (1985) has made this point clear theoretically. Without excess, there would be no destruction, without destruction there would be no creation. Last year's products must become superfluous if this year's production line is going to be sold onto the same hypersaturated markets. Consumer capitalism in particular is founded on a political economy of waste (also see Yearley, 1996).

Waste-risks pose huge challenges to the autopoietic tendencies of the main operating systems of late-modern society (Van Loon and Sabelis, 1997) and thus engender a turning in the technological culture of the risk society. Their excessive dependence on functionality and technological performativity makes waste-risks a never-ending source of potential disruption, for example as embodied by the catastrophic enthralment and anti-humanism of the promise of full autopoiesis under nanotechnology. Hence, it is no surprise that it is greeted with scepticism and even scorn, for there are no grounds for good faith within technological culture.

What is often implied in such a technological culture is a notion of time as a manipulable variable. That is, the time it takes for a system to adapt to an intervention is taken to be manageable. Technologies of recycling are designed to speed up the reproduction process by intervening in the metabolism of the social formation. The time it takes for waste to become a primary resource (say compost) is shortened by technological interventions which are required for waste to become attuned to the tempo and logic of industrial production.

Recycling is the final resort of autopoiesis. The speeding up of the processing of waste into a secondary resource is necessary for such a recycling industry to form a viable and competitive economic basis (W. Marx, 1971: 81). This means that the timescales of recycling are to become shorter and shorter. Conversely, the interval within which matter and energy take the form of waste also shrinks. Ultimately, one could argue, the interval will reach the limit of zero, by that time there is no longer any waste. This is when the acceleration and increased productivity have created a perfect autopoietic closure when the recycling process is fully self-enclosed. The perfect recycling chain could also be achieved by the complete reappropriation of all matters and energies that were 'discarded' in the primary process, that is when there are no leaks, emissions or disposals. Of course this needs to be repeated for all subsequent processes as well. This too could become an autopoietic take-off at the moment when no extra matter and energy is necessary from the outside and the system has become completely self-enclosed. In a nutshell, this is what nanotechnology promises.

Such utopian thought experiments, however, are possible only within a metaphysics of presence in which we can acquire full excess to 'the real' via the phenomena we consciously encounter *as* purely real. What such a metaphysics inhibits is a conception of a beyond, of traces, of unknown consequences, of the hidden, the repressed, the invisible and the different. The time it takes 'to come into being' is always set at the interval of zero, being is immediate, just like the present, which is presented to us as presence (Game, 1991; Van Loon, 1996a).

In a process of increased velocity and self-enclosure, the interval between emergence and disappearance of waste decreases (cf. Virilio, 1986). This is the crux of the autopoietic mode of cultivation that characterizes the biopolitics of technological culture. In its attempt to enclose all waste circulation into a singular flow under continuous acceleration, waste management generates an environment that is internal to itself. The fixation of waste in a metaphysics of presence is always a matter of mapping it between the 'not-yet' (its emergence) and the 'no-longer' (its disappearance). Waste is thus always displaced to another time-frame. Shortening the interval between the two might lead to a dissolution of waste from the present/presence, it cannot stop it from returning. To put it more simply, speeding up the processing of waste, just as exhausting waste of itself, is a mythical dissolution of the waste problem, if only because it requires increasing amounts of energy/matter input to attain those limit conditions. At the ultimate limit condition of zero-waste, all matter has become energy.

What remains outside, that which does not belong, matter out of place, non-linear risks, waste, however, will not remain 'just' a shadow on the wall, marking the boundaries of the modern, undivided, self. The embodiment of waste marks the failure of autopoiesis. AIDS, Ebola and BSE/vCJD are all deadly diseases and non-linear hazards that have emerged in the shadows under the mushroom of modernity; no doubt they have been technologically induced by 'progress': highways, hospitals, pesticides and food-processing techniques (see Chapter 7). Waste is a virulent abject of modernity, one which marks the failures of self-valorizing immune systems such as sustainable development. The abject is a concept coined by Kristeva (1988) to describe the phenomenon of an ambivalent relationship with difference. For example, if our immune system encounters a foreign body within our bloodstream, it will seek to eliminate it, to remove it. The immune system thus appropriates a distinction between self and other (Haraway, 1991; also see Chapter 4). The problem with the abject is that the otherness is generated from within. In order for the self to remain integral, its immune system has to eradicate the otherness, yet – at the same time – this becomes a violation of its own integrity, and thus a movement of self-destruction.

Virulence adds something to the abject, namely that it becomes self-reproductive, usually in close (parasitical) association with the system to which it relates as an abject. Virulent abjectivity is like a degenerating auto immune deficiency, it produces more and more of itself because the immune system is geared towards its expulsion yet – simultaneously – destroys itself. Virulent abjectivity is not only confined to the world of viruses, but also applies to radiation, genetic modification, biochemical toxins, biological warfare and 'global terrorism'. All are forms of contagion that are aggressive, predatory, parasitical and irreversible.

7 Emergent pathogen virulence

Understanding epidemics in apocalypse culture

The history of our time will be marked by recurrent eruptions of newly discovered diseases . . . ; epidemics of diseases migrating to new areas . . . ; diseases which become important through human technologies . . . ; and diseases which spring from insects and animals to humans, through man-made disruptions in local habitats. To some extent, each of these processes has been occurring throughout history. What is new, however, is the increased potential that at least some of these diseases will generate large-scale, even world-wide epidemics.

(Garrett, 1994: xv)

This chapter focuses on the problematic of emergent pathogen virulence of viruses, bacteria and other pathogens. Using Ebola as an example, the aim is to show how emergent virulence is endemic to the way in which 'modernity' has evolved. Furthermore, the objective of this exploration is to trace the way in which discussions over 'emergent viruses' intersect with the spreading of a more general apocalyptic ethos in popular culture and what kind of economic, social, political and cultural factors operate at the roots of such an intersection.

It is of little relevance whether the agents of risk are organic or inorganic; their effects both relate to processes of 'contamination' and 'spreading'. They can both be understood as 'actors'. However, what may set them apart and force us to rethink the conceptual framework outlined in part one is the question of 'motivation'. In the discourses that have brought viruses to our attention, pathogen motivation is of crucial importance. Viruses make us ill because they are replicating themselves; like waste, they are virulent abjects of modernity. However, unlike waste, they 'take over' bits and pieces of our bodies because they are motivated by self-replication. That is, they borrow bits of genetic material (DNA or RNA) and ribosomes from their hosts (Cann, 1997; Levine, 1992). In nuclear physics, discourses of radiation do not address the issue of motivation. A radioactive particle is not motivated to engage in self-replication. Although we may conceive of radiation as itself a form of motivation, for example as matter-becoming-energy; this motivation is not part of a strategy to replicate itself; its logic is exterior to its own movement (transcendent). Viruses, in contrast, evolve with an immanent motivation alongside a transcendent one.

In 1996 the WHO reported that:

> Far from being over, the struggle to control infectious diseases has become increasingly difficult. Diseases that seemed to be subdued, such as tuberculosis and malaria, are fighting back with renewed ferocity. Some, such as cholera and yellow fever, are striking in regions once thought safe from them. Other infections are now so resistant to drugs that they are virtually untreatable. In addition, deadly new diseases such as Ebola haemorrhagic fever, for which there is no cure or vaccine, are emerging in many parts of the world. At the same time, the sinister role of hepatitis viruses and other infectious agents in the development of many types of cancer is becoming increasingly evident. The result amounts to a global crisis: no country is safe from infectious diseases. The socio-economic development of many countries is being crippled by the burden of these diseases. Much of the progress achieved in recent decades towards improving human health is now at risk.

> (World Health Organization, 1996: 1)

Newly emergent viruses such as AIDS and Ebola, the return of old and nearly forgotten lethal disease such as tuberculosis, cholera and the bubonic plague, a dramatic reduction in the effectiveness of antibiotics, the rapid emergence of so-called superbugs, and the arrival of entirely mysterious and unknown non-living but equally lethal pathogens such as prions, are all signs of the return of a repressed. Whereas in the 1970s, medical scientists and politicians claimed victory in the war against infectious disease, the emerging picture is quite the reverse. Parasites, bacteria, viruses and prions are gaining the upper hand (see for example Cannon, 1995; Garrett, 1994; Morse, 1993; Ryan, 1996). Even evolutionary biologists such as Ewald (1994) and Wills (1996), whose views are less apocalyptic because they see the human as the most successfully evolved species, have conceded that the rise of emergent pathogen virulence is alarming.

In the previous chapter, it became apparent that ecological risks such as those related to waste and pollution force us to think about 'social order' in quite different ways than those professed by traditional sociology and more generally the episteme of 'modern thought', with its binary oppositions between individual and society, structure and agency, reality and representation, universal and particular, mind and body and nature and culture. In particular with reference to the last division, we shall see that emergent pathogen virulence is not simply a 'natural' phenomenon.

The image of the looming resurgence of pandemic plagues of infectious diseases illustrates Ulrich Beck's (1992) risk society thesis. What is obvious in applying Beck's thought to the problematic of waste and waste management is the 'manufactured' *nature* of risk and uncertainty. That is, the way in which industrial capitalism operates finds an unforeseen side-effect in the production of unwanted elements, which in turn are being reprocessed back into the production system. Indeed, the rise of the recycling industry shows that waste has become that which 'valorizes' commodities and informs 'use value'. Consequently, it also shows that understanding 'risk' as 'manufactured uncertainty' is not in itself sufficient to make sense of waste; we should also talk about risk *management* as both comprising of strategies of risk containment and risk aversion *and* of the transformation of risk into opportunities, for example through valorization.

The leading question of this chapter builds upon this insight and asks how to conceptualize 'health risks' in particular those related to infectious diseases. Like waste-risks, the increase in infectious diseases can be theorized from the perspective of manufactured uncertainties. However, if we are solely focused on the manufactured risks of infectious disease, we may neglect the ways in which the pathogens themselves are being enrolled into a range of actor networks. That is to say, pathogens too may enact intelligent responses to the human drive to modify its environment. This has particularly significant consequences for the way in which 'epidemic management' is to be conceptualized and implemented. Far from being a set of terms and practices aimed at the objectification of an 'alien threat', epidemic management needs to engage reflexively with the way in which it is caught up in the reproduction of risks and hazards.

The chapter starts with one example of emergent pathogen virulence: Ebola. This sets up the argument that the sudden emergence of Ebola as a highly lethal and infectious disease resonates acutely well with cultural sentiments that were already prevalent in modern western technological culture. Such resonances, however, are themselves framed in globally disseminated political, economic, social and cultural constellations that have enabled scientists and media pundits to refer to emergent pathogen virulence as 'nature's revenge'. Two examples will illustrate this further: the cultural significance of HIV/AIDS and an emergent genre of 'epidemiological science/fiction', exemplified by Michael Crichton's *The Andromeda Strain* of 1969.

The previous chapters highlighted the importance of cultural factors, not only in the interpretation of 'risks' and 'hazards' in the context of modernization, but also in their creation, and more specifically, the wider technological culture in which they emerge. Discussions over infectious diseases and epidemics are no exception. They are placed at the core of the way in which we perceive and manage the predicament of the human species. Indeed, accounts of the increase in pathogen virulence have resonated particularly well with a growing uncertainty over the future of our technological culture. Viruses are the messengers of the end of the world as we know it. Therefore, the ultimate objective of the following analysis is to focus more directly on the way in which medical science and popular culture have 'collaborated' on the production of sense and sensibility of pathogen virulence.

Popular culture can indeed be seen as playing a crucial role in the social and symbolic organization of risk management; expositions of newly emergent pathogen virulence have fully embraced the technological culture of the risk society. However, our exploration would not be able to escape the ironic turning-inward if it would merely circulate on the plane of textual analysis. Therefore, we turn to more sociological explanations of infections and epidemics to argue that pathogen virulence is part of a wider network of actors (humans, animals, technologies, spirits). Moreover, it allows us to understand the social in terms of a complex spatialization of body politics and biopolitics, in which pathogen virulence constitutes a particularly effective medium of both 'sense-making' and the management of body boundaries.

Emergent pathogen virulence: Ebola

When in August 1976, a lethal haemorrhagic disease emerged in the missionary hospital of Yambuku, a small village on the Ebola river in Zaire, it struck with a vengeance reminiscent of the 'Black Death' that raged throughout Europe during the fourteenth and fifteenth centuries. In total 277 people were infected, of whom 257 died. With a mortality rate of almost 93 per cent, Ebola Zaire, as the disease would later be called, would go down in medical history as one of the most lethal infectious diseases ever known (Johnson, 1982). In this and later cases of outbreaks of Ebola, it was the medical system itself which played a major role in engendering the epidemic. Syringe transmission was identified as one of the key vectors in the spreading of the disease, which, together with the concentration of people and poor hygienic environments, turned medical centres such as clinics and hospitals into 'hot zones' of infection (Peters *et al.*, 1993: 162).

> One room in the hospital had not been cleaned up. No one, not even the nuns, had had the courage to enter the obstetrics ward ... The room had been abandoned in the middle of childbirths, where dying mothers had aborted foetuses infected with Ebola. The team had discovered the red chamber of the virus queen at the end of the earth, where the life form had amplified through mothers and their unborn children.
>
> (Preston, 1995: 133)

Of the newly emergent viruses, there are few that have been able to mobilize such a fear and fascination as Ebola (Preston, 1995; Ungar, 1998). Ebola is often mentioned as the most gruesome example of nature's revenge (Preston, 1995). It also sparked off a short-lived but hyper-intensive media panic (Ungar, 1998). A few factors may have been involved in Ebola's appeal to the popular imagination (Ryan, 1996):

- the death rate (60 per cent for Ebola Sudan and 90 per cent for Ebola Zaire)
- the speed at which the disease spread (the incubation is between seven and ten days)
- the gruesome way in which people are being killed by the virus (literally bleeding to death)
- the absence of a cure

- the total enigma that the virus posed in terms of its identity and its natural origin.

Ebola is indeed a very mysterious disease. Its obscurity has been one of the main obstacles towards its objectification; even if the aetiology of the disease is quite well known, its origin is not. What is known about Ebola is that it is a 'filovirus', which as a viral family is virtually unknown. The first recorded human encounter with a filovirus was in 1967 in a laboratory in Marburg, Germany. In total seven laboratory workers died from initial contact with this virus and twenty-four others were infected (Peters *et al.*, 1993: 159–61). Ebola was next in line to be identified as a filovirus. In 1976, it emerged in two places at the same time, in Zaire and Sudan, a strange coincidence as the diseases turned out to be of two different strains. The name Ebola was derived from the Ebola River, close to where the missionary hospital in Yambuku was located. The disease was more lethal than Marburg and caused widespread concern in the medical establishments across the world. In both cases hospitals and dirty needles were identified as the major vectors of transmission.

On the basis of the subsequent collection of blood samples, virologists were able to identify the Ebola virus as being a string of seven RNA nuclei. RNA viruses are known for their rapid evolution rates and flexible adaptability. They consist of extremely heterogeneous populations (Holland, 1993: 204). Ebola attacks cells and in particular thrombocytes (blood platelets) and endothelial cells. It stops the blood from clotting. The overall effect is that infected organisms develop internal bleedings that do not stop. Richard Preston, in his dramatic account of the final seconds of someone dying from such a filovirus, describes this in terms of 'crashing':

> He becomes dizzy and utterly weak, and his spine goes limp and nerveless and he loses all balance. The room is turning round and round. He is going into shock. He is crashing. He can't stop it. He leans over, head on his knees, and brings up an incredible quantity of blood from his stomach and spills it onto the floor with a grasping groan. He loses consciousness and pitches forward onto the floor. The only sound is a choking in his throat as he continues to vomit while unconscious. Then comes a sound like a bed-sheet being torn in half, which is the sound of his bowels opening at the sphincter and venting blood. This blood is mixed with intestinal lining. He has sloughed his gut.
>
> (Preston, 1995: 51)

McCormick and Fisher-Hoch (1996: 168) however claim that Preston's dramatic version of the workings of Ebola misrepresents the way in which the disease operates. Instead of 'liquefying' all forms of organic tissue, Ebola attacks only the blood and blood-vessel cells; as a result blood starts leaking out of the body. The point to make here, however, is that there is no scientific closure on what Ebola does. This is not only because of a lack of sustained scientific research, but also because Ebola operates as a cultural enigma, an alien signifier, an exteriority that disrupts the smooth functioning of the abstract machine. Whereas the medical-scientific establishment has been developing some under-standing of the form and operational logic of Ebola, there is still nothing known about the natural habitat of the disease. Epidemiologists and virologist hunting for the 'origin' of Ebola have literally as well as metaphorically run into dead ends. The disease vanishes with its last victims. In most epidemic outbreaks, the 'index case' could not be traced. Moreover, the natural 'host' of these and other filoviruses remains unknown, although there are some indications that the Marburg virus (the first known filovirus) is resident in green monkeys (Peters et al., 1993: 163). Hence, a genealogy of filoviruses has not yet developed. There is no well-supported evidence of any origin. In absence of any 'scientific evidence', commentators refer to the general common denominator of 'Nature'.

Because of their sporadic occurrence, and perhaps the relative low threat they pose to western and in particular North American society, interest in the Ebola virus has dropped considerably. It has not been able to sustain its enrolment of an effective assemblage upon the political, economic, social and cultural infra-structures of modern society. Its appeal to a public concern might have been intense, but it did not last very long. Apart from a high intensity and sense of urgency, effective and sustainable actor networks also require a frequency and durability of concern.

Relatively recent developments have not changed that. In 1989 and 1990, two successive outbreaks of Ebola among monkeys in Reston, Virginia should perhaps have sparked off a renewed public interest in the disease, in particular because this time the Ebola virus seem to have been transmitted via aerosol routes, but it hardly made any headlines. However, although four people were infected (they did not become very ill), the disease wiped out a near 100 per cent of the primate population. In 1995, 1996 and 1997 there were successive outbreaks of Ebola in Kikwit (Zaire) and two in Gabon respectively. Finally, there was a major outbreak of Ebola in Uganda in 2000. The 1995 Kikwit outbreak proved to be the worst recorded case in history, with 315 infected, of whom 77 per cent died. It also captured the attention of the media, as stories poured out of a new plague type of disease for which there was no cure. These stories were often accompanied by claims that Ebola had arrived in the western world, whenever a plane landed in a European or North American airport with someone from Zaire on board (Ungar, 1998).

Although the hype around Ebola has died down, it now regularly features alongside AIDS in the growing lists of coming plagues. Also featuring are, for example, Hantaviruses (one of which claimed the lives of many young people during an outbreak in New Mexico in 1993), Hepatitis viruses, haemorrhagic fever viruses such as Machupo, Junin (both South America), Kyasanu Forest (India) and Lassa (West Africa; also see Fuller, 1974), various encephalitis viruses and various strains of the influenza viruses.[1] Many of these diseases are highly contagious and can be lethal, although the severity varies. Morse (1993: 15) argues that most of these 'emergent viruses' are not new but simply existing viruses that have evolved to conquer new territory. He mentions in particular diseases whose natural hosts and primary vectors are animals (zoo-notic viruses). As humans are expanding into the spaces previously occupied by for example rodents or monkeys, there is more opportunity for inter-species viral traffic. In this sense, their emergence is intricately connected to the behaviour of humans and its impact on the global ecosystem.

Explanations of increased pathogen virulence

The growing list of newly emergent pathogens such as Ebola and returning old ones (i.e. cholera, plague) raises two distinct questions. First, it forces us to ask what causes each of these diseases to 'emerge'? Here, no general answers can be given, as each disease has its own multiple aetiology, many of which subject to pure speculation. However, at the same time, we may ask what causes the growing of the list as such? Here, answers are sought that supersede each individual case and provide the logic of the emergence of a discourse on 'emergent viruses'.

However, one may dispute the very claim that 'emergent viruses' are a serious problem. The only reason why the list has increased so rapidly since the early 1970s is that – alongside developments in medical virology and molecular biology – diagnostic tools for the identification of new viruses have vastly improved (Waterson and Wilkinson, 1978). Rather than a sign of a doom, it should be celebrated since it signals that we now know more about viruses than ever before in history. Alongside the 'improved diagnostics' argument, a similar dismissive explanation could be that we simply encounter more unknown viruses now than in the past because we are in closer contact with the most remote parts of the world. Whereas up to the early 1940s, epidemics could affect large populations of the world without being detected by western public health surveillance systems, this is no longer the case. Under globalization (spear-headed by the WHO), no epidemic will remain a secret.

Both aforementioned explanations share a similar dismissal of the problem. There is no increase in emergent pathogen virulence, only an increase in information and knowledge about them. The question why there is so much interest in this issue can be equally dismissed. The ability and willingness to tap into a discourse of emergent pathogen virulence relates to other (not specifically medical) factors such as competition for status, limited funds or sponsorship from the pharmaceutical or military industrial complex. People who believe this will undoubtedly sleep much better than those who take the phenomenon of emergent viruses more seriously.

The second type of explanation belongs to those who see emergent pathogen virulence as something 'real', i.e. something caused by factors outside the field of epidemiology. There are numerous indications that ecological, economic, political, social, cultural and technological forces may have contributed to emergent pathogen virulence. Some of these factors could also be used to explain why emergent pathogen virulence has become a major public health concern. That is, not only do they point towards motivations operating in subcellular assemblages, but also they extend into other territorialities, particularly those dominated by technoscience, governance, media, commerce and the military.

From our discussions of risk in Part I, we can deduce that risk is a virtual object, a becoming-real. That is to say, it has no existence outside sensibility and perception. Hence, whether one is more inclined to side with a dismissive or affirmative response to the question whether or not emergent pathogen virulence is 'real', the most important aspect remains that its discourse is real enough for public health authorities to invest considerable amounts of resources and credit into increased security and regulation, not to mention the commercial interests in pharmaceutics as well as entertainment. That is, as a risk, emergent pathogen virulence has become real enough. In this becoming-real of risk, a focus on 'exogenous' factors as playing a major role becomes inevitable. Central to this is the allusion to the generic 'emergence' of a discourse of emergent viruses. The analysis concentrates on possible factors that have caused increased pathogen virulence as if it represents an 'objective' reality.

The first set of causes of emergent pathogen virulence are found in the global ecology. Deforestation, erosion of the soil, the depletion of the ozone layer, climate changes, pollution of waters, air and land, acid rain, the extinction of many species of animals and plants have all been related to emergent pathogen virulence (see, for example, Ryan, 1996; Preston, 1995). The underlying argument is that under the culmination of ecological disasters, many microorganisms are losing their natural habitat, for example because their natural hosts are becoming extinct. The organisms may mutate or seize new opportunities – and thus increasingly enter our human habitats. As we humans expand our activities, we enter those habitats and contribute to an increased traffic of pathogen virulence between species.

Christopher Wills (1996) makes a very strong case for the argument that understanding how human diseases evolve requires not simply an analysis of the interaction between humans and pathogens, but also an account of the intermediary roles of animals and parasites. In his excellent analysis of the development of the Black Death in Europe, for example, he incorporates four vectors: humans, animals (rats as well as pets and cattle), bacteria and fleas. Only by incorporating all four does it become possible to understand why the disease could have such devastating effects on medieval Europe and how it could spread despite measures being taken that would seem adequate. More specifically, he is able to show how a rather untidy and far-from-smooth series of genetic mutations has played a central role not only in the evolution of the disease, but also in its subsequent disappearance. The relationship between these four elements, however, constitutes only a part of the actor networks of emergent pathogen virulence.

Viruses and bacteria are extremely adaptable creatures and they mutate quickly to adjust themselves to new environments. Climate changes, radiation, increased transformation of energy, biochemical toxicity might all contribute to such mutations and may result in the formation of new, aggressive, parasites. As

McNeill (1976/1994) has shown, the history of human civilization is also a story of the changing associations between human hosts and microbiotic organisms. Following Symbiosis Theory (Margulis, 1993; Lederberg, 1993), Ryan (1996) suggests that such an evolution by association (Sapp, 1994) has been by and large mutually beneficial to both host and parasite, even if the symbiosis is an aggressive one.

Viral symbiosis, however, has a peculiar quality. It takes place on the level of genetic coding. It changes the genetic coding of all of its hosts; those whom it protects as well as those whom it destroys. This points towards a 'genomic intelligence' that strikes at the core of understanding the politics of survival in terms of the ability to modify one's environment.

> It is hardly accidental that the increased emergence of viruses, and other microbes, affecting people in the second half of this century has paralleled the extent of human invasion of wilderness ecologies such as the rainforests and of massive alteration of savannah and other similar ecologies through farming.
>
> (Ryan, 1996: 285)

Ryan uses the concept of junctural zone to refer to the space in which two different species meet, which are not too dissimilar but have not had much previous contact either. This is the zone in which an aggressive symbiont might appear. HIV, Ebola, Hanta and Lassa viruses can all be seen as aggressive symbionts that have emerged in the junctural zones between humans and primates, and humans and rodents respectively. The junctural zone, one could argue, is that liminal space where difference matters. The junctural zone is thus a crucial space, a boundary or borderland, that marks the territoriality of order as distinct from that of disorder (the jungle). The junctural zone is the space of transformation, the space of warfare, indeed the space where the politics of survival can be pursued in its most aggressive form.

The second set of causes are often related to the first, but have a more specifically social nature. McNeill (1976/1994) has convincingly argued that there is a symbiotic relationship between warfare and epidemics as the conquering forces often bring with them pathogen virulence against which the conquered have no resistance. The strength of the conquering also depends on their ability to develop an endemic relationship with potentially lethal bacteria and viruses. Conquered peoples often do not have similar symbiotic relationships because they lack the mobility, speed and concentration in numbers to develop sufficient immunity.

This also relates to industrial projects such as motorways (McKechnie and Welsh, 1993; Preston, 1995),[2] dams (Shope and Evans, 1993) and urbanization (McCormick and Fisher-Hoch, 1996).[3] Apart from a direct ecological impact, they also facilitate specific concentrations of human beings, animals, parasites and pathogens, for example in the case of water reservoirs which enable the rapid reproduction of arthropod vectors such as mosquitoes, which themselves function as 'reservoirs' for the rapid accumulation of pathogens such as malaria.

Further changes in social behaviour, such as sexual codes and kinship relations, often related to the uprooting of traditional social structures by 'modernization', are seen as essential components of the successful spread of diseases such as AIDS, hepatitis and meningitis. For example, women in urbanized Africa often lack the support of husbands and families that they get in the more rural areas. They often make a living as 'free women', which is not exactly the same as prostitution, but exposes them to many of the same risks and dangers as prostitutes (McCormick and Fisher-Hoch, 1996). Furthermore, AIDS, for example, is often associated with migration (Haour-Knipe and Rector, 1996), and although much of this association is culturally engendered (e.g. stigmatization: Browning, 1998; Patton, 1990; Treichler, 1988; Waldby, 1996), there is an obvious connection between the large-scale movement of people (which includes refugees as well as business travel and tourism) and the spreading of diseases.

A third set of factors causing pathogen virulence to emerge on an increasing scale are related to the way in which such virulence is managed. A telling example is the fear not only that the effectiveness of antibiotics is decreasing, but also that it contributes to an intensification of bacterial virulence (Ewald, 1994). This phenomenon has caused widespread concern over the way in which we have become so highly dependent upon the medical technoscientific 'fixes' (Cannon, 1995; Van Loon, 1997a). At the same time, calls for a more effective global strategy in dealing with this problem have been systematically ignored by pharmaceutical industries, which – in an environment of global competitiveness and market economics – are forced to protect their corporate interests rather than their own corporeal ones (Garrett, 1994).

Equally distressing is the rise in possible forms of transmissible spongiform encephalopathies and their illustrious agent, which is assumed to be the prion, among various species of animals, including humans; this is generally linked to the industrialization and mass production of food as well as to the way in which agricultural business is forced to operate in a climate of global, subsidized pseudo-market competitiveness (Ford, 1996; Macnaghten and Urry, 1998; Toolis, 2001; Van Loon, 2002). According to Toolis (2001: 23), ever since the first utterances were made about potential links between BSE and vCJD, 'every arm of the British state . . . colluded in a conspiracy of denial' making BSE

> probably the most cynical act of biological warfare ever waged against a civilian population by a western government. The British government put the narrow business interests of its farming and meat industries before the health of its population and that of other countries.

> (Toolis, 2001: 27)

Indeed, the mere fact that we are unable to distinguish between the ostrich-politics of secrecy and denial and the evil cynicism of party politics mixed with capitalist greed, indicates that in the face of risks such as emergent pathogen virulence, we can no longer find comfort in either knowledge or ignorance, secrecy or publicness. As technoscience, governance, mediascapes, commerce, law and military powers are increasingly intertwined, a radical doubt or, to speak with Kant, a 'radical evil' (Copjec, 1996) becomes equally widespread. It is therefore necessary to deconstruct the intersystemic links that constitute the technocultural organization of health risks, in order to understand how such risks are generated from within these constellations. The actor networks involved in the management and prevention of epidemic risks are far less coherent than is generally claimed. The multiplicity and paradoxicality of the risks involved further adds to a growing awareness that the networks need to become more responsive to political, social and cultural issues.

At the same time, we can see that the actor networks of the risk society are losing their tight grip on the social and symbolic organization of our technological culture. In the face of biotechnology, genetics, and (as we shall see in the next chapter) cybernetics, the established modes of operation of these networks are simply inadequate to deal with the challenges. What we encounter is the crisis of crisis management. Even if there are no military or industrial conspiracies that have a stake in creating pathogen virulence (and this is certainly not to be dismissed), we can see that the existent global-capitalist system will never be able to respond to the perceived risks adequately. There are economic as well as political stakes in protecting the global asymmetry of health risks. Indeed, whereas the agents of epidemic risks are micro-organisms, and their vectors often other non-human life forms, the irregularities in their emergence and spreading cannot be dissociated from processes of modernization and industrialization.

As we have seen in previous chapters, risk equals opportunities and profitable business ventures. As the example of AIDS illustrates, the pharmaceutical industry in particular has played a major role in the transformation of epidemic risks into profitable opportunities. The consequences, however, are often counterproductive. For example, the institutionalized pharmaco-logic of (medical) science (including the one applied in the meat industry) cannot be ignored in understanding the decreasing effectivity of antibiotics. Also, the role of the military-industrial complex, not only in researching but also in developing pathogen virulence, must not be ignored, as it is here where the most innovative technoscientific research (with all its inherent risks and hazards) is likely to take place. As Virilio (1986) has noted, almost all technological innovations have an origin in warfare (also see Deleuze and Guattari, 1988). This also applies to microbiology. The link with the possible interference of military motivations in the struggle against emergent pathogen virulence of course directly resonates with a more popular sentiment, that there is a more sinister and diabolic 'evil' operating behind the scenes of this medico-scientific spectacle; one which has very little to do with the motivation of the micro-organisms themselves, but more so with megalomaniac desires of mad science, bad business and evil warlords.

Signs of the times: an emergent apocalypse culture

Since 1992, alarm over emerging and re-emerging diseases has resulted in a number of national and international initiatives to restore and improve surveillance and control of communicable diseases. In 1995, a resolution of the World Health Assembly (WHA) urged all Member States to strengthen surveillance for infectious diseases in order to promptly detect re-emerging diseases and identify new infectious diseases. This resolution led to WHO's establishment of the Division of Emerging and other Communicable Diseases Surveillance and Control (EMC), whose mission is to strengthen national and international capacity in the surveillance and control of communicable diseases, including those that represent new, emerging and re-emerging public health problems.

(WHO, 1998)

The growing concern over emergent pathogen virulence addresses both the general increase in the number and diversity of epidemics as well as the increased awareness that the powers of medical technoscience to combat such outbreaks are rather limited. This concern has now also affected (or infected) the domains of popular science and science fiction writings which have disseminated this sense of urgency over emergent viruses towards a wider audience. Although the balance between containment and moral panic is still favourably biased towards the first (Ungar, 1998), the relatively recent out-pouring of writings on new and virtually unknown diseases signals a greater receptivity among a wider audience of the suggestion that we may be at the brink of a series of pandemics which resemble those of the medieval plagues.

For example, two films entitled *Virus* (based on Robin Cook's medical thriller *Outbreak*) and *Outbreak* both dealt with Ebola outbreaks in the USA. Publications bordering on medical science and popular fiction such as Preston's (1995) *The Hot Zone*, McCormick and Fisher-Hoch's (1996) *Level 4: Virus Hunters of the CDC*, Robin Cook's (1987) *Outbreak*, William T. Close's (1995) *Ebola* and Frank Ryan's (1996) *Virus X* all dealt with issues surrounding Ebola. Moreover, Garrett's (1994) *The Coming Plague*, Christopher Wills' (1996) *Yellow Fever Black Goddess* and Geoffrey Cannon's (1995) *Superbug* incorporated the notion of the incurable disease in their apocalyptic accounts of the consequences of newly infectious diseases, and the return of old ones, for our future human predicament.

Many of the accounts of Ebola and similar emergent and unknown infectious diseases derive their main plot and popular appeal from Michael Crichton's acclaimed *The Andromeda Strain* which was published in 1969 and was a main trendsetter in blending medico-scientific discourse and popular-thriller narrative. What was crucial about this book was that in terms of its references and construction, the distinction between science and fiction was purposefully obliterated, disrupting the otherwise comfortable disclaimer that it is merely fiction. Its main appeal is that it could easily happen, and it showed how inadequate even the most minutely planned attempts to eradicate contingencies are. The idea that a satellite returning to earth may have picked up some strange and unknown pathogen, which is either alien or a mutation, is of course not at all far-fetched and indeed already widely discussed in worlds of science and the governance of national security. The 'little accidents' that may seem like 'bad plots' reflect the inherent connections between contingency and risk which tend to be ignored in the representations of technoscience, governance and military strategy because of their aberrant nature.

Even in the more cut-and-dried corners of serious writing, similar alarming claims about 'cultivating' increased virulence among antibiotic-resistant bacteria have equally been made. Indeed, Ewald's (1994) evolutionary biological approach to the rise of infectious diseases gives a detailed account of how an evolutionary epidemiological approach could explain the emergence of the increased pathogen virulence of Methycillin-Resistant *Staphylococcus aureus* (MRSA). His words, however more cautious, are still like those of a preacher of doom:

> Whether or not the broad antibiotic resistance is evolutionarily linked to virulence, this threat has become more ominous as the range of antibiotic resistance has broadened without a concomitant increase in the classes of effective antibiotics. And the problem is not restricted to staph. Resistant hospital-acquired bacteria have now been found across taxonomic groups and in virtually all hospitals in the United States.
>
> (Ewald, 1994: 98)

With the exception of Ewald and Wills, the aforementioned authors all write on the borderlines of popular science and popular fiction. Their appropriation of a split between anxiety/fear on the one hand, and libidinal death-drive fascinations on the other hand, must therefore be firmly placed in a 'popular culture' in which such an apocalyptic ambience 'makes sense' and has a certain appeal. Of course, science and popular culture are intimately intertwined; they are part of the same modernist neurosis that has swept large parts of the world and provided a transition from the Second World War to the Cold War. Playing the game of exchanging threats and opportunities, such science-fiction hybrids constitute a world view in which we have to rely on what we distrust (i.e. modern institutional forms) to provide the solutions by extending their impact. The resilience of microbes and viruses is not seen as a sign that the world may not be the one we once thought we had designed: 'It's life, Jim, but not as we know it' is not a far cry from the encounter many phase-4 lab technicians have had with many new sorts of viruses and microbes.

There is an astute resonance with a sense of apocalypse here as a turning in technological culture. After all, the coming plagues have been announced several times in the biblical New Testament. They signal the coming of the Anti-Christ and the end of the world as we know it. The fact that many of these writings project this emergent future in terms of a millennial anxiety – even no longer with the added appeal of the becoming-real of the Millennium Bug – further strengthens the idea that newly emergent viruses are in fact nothing less than the (diabolic) messengers of the eschaton. Against such mystical anxieties, truly modern institutions such as the WHO have taken upon themselves the formidable task of managing the risks to public health that such an ensemble of pathogen virulence may bring upon humanity as a whole. Indeed, while many of the aforementioned publications do ride on the waves of the apocalypse, most of them end with a more optimistic message – that it is still not too late to do something about it. Technological culture can still save us, if only we were able to control some of the excessive contingencies such as bad faith, selfishness, greed, pride and ignorance.

On the intersections between technoscience and popular culture – that is technological culture – where sense is made, indeed where pathogen virulence is made intelligible, we thus trace a close encounter of another kind: between the religious eschatology of the apocalypse, an immanent nihilism and the modernist teleology of progress. In order to make some sense of this strange encounter, we will genealogically trace the emergence of writings on emergent viruses to two 'origins' – one which is the discovery of HIV in medical science, the other is the aforementioned piece of literary science fiction: Michael Crichton's *The Andromeda Strain*.

HIV/AIDS

> If HIV did cause AIDS it would be one of the smartest viruses you could imagine, because it knows whether you're male or female, gay or straight, white or black, it even knows what zip code you live in and what country you live in. For example, in the United States eight out of nine AIDS cases are men. In San Francisco 99 out of 100 are men. Two-thirds of AIDS cases in the USA and Europe are gay men. So that's a smart virus, and somehow that virus knows to cause Kaposi's sarcoma in gay men and not in hemophiliacs, not in pediatric AIDS cases, not in IV, blood transfusion or so forth. So it's a very smart virus.
>
> (Rasnick, 2000)

The arrival of AIDS as a global pandemic has caused widespread concern, not only within the medical science and public health establishments, but also within the wider realms of popular culture and politics (Oppenheimer and Rickitt, 1987; Shilts, 1987). In the language of Actor Network Theory, the virus has been able to enrol a worldwide network of actants to a degree that some critical commentators talk of an 'AIDS establishment' that conspires against any criticism of the ruling claims about the cause and extent of the pandemic (Ankomah, 1998; Rasnick, 2000). Frank Ryan (1996) captures the popular sensibility towards AIDS really well in the following passage:

> AIDS was a startling vision: an emerging virus, contracted in the main by sex amongst the young, a subtle and sinister infection that kills the very cells that might fight it. This destruction of the immunity of the victim allows all manner of secondary horrors to invade, so people drown in

amoeba, MAC germs from tap water invade their bowels and enter every crevice of the internal organs, strange viruses, such as the cytomegalovirus, hop a ride. Their skins are showered with purple-bluish cancers. Bizarre forms of leukaemia course through their blood streams. Every surface of the body, both skin and internal, is prey to a myriad of disfiguring, debilitating and tormenting afflictions, so that the sufferer is compelled to take a night-marish cocktail of relatively toxic drugs just to try and keep down some of these secondary manifestations.

(Ryan, 1996: 243–4)

It is the combination of the unknown nature of the virus (some even question whether it is HIV that causes AIDS), its lethal effects, its appropriation of particular vectors, in particular those of sex and drugs, its global spread and the absence of any known cure (Mann *et al.*, 1992: 16–19) that have made AIDS the archetypical apocalyptic disease. AIDS has affected the very core of the western-dominated and globally present cultural formations of individualism and risk perception. However, it is the specific relationship between AIDS and metaphor (Sontag, 1989) that may be seen as providing the most imaginative and revealing analysis of its impact on contemporary society (Browning, 1998; Lupton, 1994; Pastore, 1993; Patton, 1990; Treichler, 1988; Waldby, 1996). Indeed, with the proliferation of diseases associated with AIDS and HIV, we also encounter a spreading of metaphors and symbolic associations that hover between hypermodern narratives of space-invaders and archaic meditations on God's vengeful wrath on the human being. Of course, as Patton (1990: 2) suggests, we *must* deploy metaphors to make sense of the fundamental unintelligibility of the disease in our everyday experience; yet at the same time, this 'must' is deeply political and thus subject to discursive struggle and negotiation.

Indeed, there is a lot more to AIDS than its spreading through HIV. It can be, and has been, politicized in social and cultural practices to engender forms of knowledge that are attuned to the sensibilities of people who are not part of the AIDS establishment. Barbara Browning describes the violent associations 'between the AIDS pandemic and African diasporic cultures' (1998: 7). In a similar vein, John Fiske (1994: 191) cites Zears Miles from Black Liberation Radio who suggested that AIDS has been engineered by the biowarfare industry of the US government as an instrument of genocide. Although the politics in both examples are quite different, the association between the disease and acts of violence is similar: they reconfigure the microbiological world within a framework of macrobiology and biopolitics (also see Bounan, 1991).

The mobilizing effects of the cultivation of a public sense-making and sensibility of AIDS thus dragged the virus out of the laboratories and test tubes of medical science and out of the specialist care of hospitals and medical institutions into the public sphere. AIDS entered popular culture in 1980s and subsequently transformed the way in which infectious disease operated in cultural politics (Oppenheimer, 1997; Paillard, 1998; Waldby, 1996). From here it is only a small step to the development of a more general public concern over emergent viruses. AIDS provided merely the index case – a new disease that was unknown, lethal, contagious and globally present.

However, this in itself does not entirely explain how emergent viruses and especially HIV/AIDS have come to play such a dominant role in contemporary cultural images of the apocalypse. For example, what allowed the cultivation of popular imagery of newly emergent viruses to connect so smoothly to an apoc- alyptic sensibility that the disease is either seen as the product of God's will or of a sinister conspiracy against human being? Simply referring to millennial anxiety is not enough, since what is so peculiar about this anxiety is that it is not based on mythical or psychological meditations on the meaning of disease (as was the case, for example, with tuberculosis and syphilis which had a certain 'romantic', 'bohemian' or 'gothic' appeal in the nineteenth century), but instead on a widespread, detailed and fairly accurate understanding of the microbiolo- gical processes that underscored the narrative of the transformation from HIV infection to full-blown AIDS (e.g. Wills, 1996). That is to say, the apocalyptic sensibility that emerged under the mushrooms of AIDS was far from 'irra- tional', but written in the very same texture as that of medical science.

Crichton's The Andromeda Strain

When reflecting on the intersections between science and popular culture, we should also look at the way in which the latter has influenced the former. When the film *Jurassic Park* (directed by Steven Spielberg) made film history in 1993, it introduced the topic of genetic engineering to a mass audience. In this film, the dramatic consequences of human 'interference with nature' were ruthlessly (if humorously) exploited by the raptors who once being able to 'mate' (due to a very obvious negligence on behalf of the scientists) had no trouble in restoring a more Jurassic political order. The film had such a strong appeal that it forced those involved in the Human Genome Project to state in their PR campaigns that their involvement in genetics was not like that of *Jurassic Park*, and indeed that genetic science was nothing like the popular image of it promoted by *Jurassic Park*. Like Mary Shelley's *Frankenstein*, the image of 'mad science' as depicted in this film had an immense influence on popular understanding of science against which the scientific establishment has battled without much effect.

The script was based on a book by the same title written by Michael Crichton. Crichton was already a well-established name in science-fictional writing. One of his most famous books was written in 1969 – *The Andromeda Strain*. Immediately striking about this book is its usage of diagrams, detailed footnotes and references, which is unusual for any piece of fiction. The distinc- tions between science and fiction are being blurred from the outset.[4] Crichton goes to great lengths to suggest that there is a thoroughly scientific case for this narrative. And there is, of course, for what he is talking about – the possibility for introducing new life forms on earth through the effects of space explorations – is not that far-fetched, certainly if we include possible mutations of earthly life forms. This story deals with the possible consequences of such an event and focuses on the practices of 'discovery' that take place through interactions between different scientists in a military-industrial laboratory setting. He shows that even the most sophisticated systems of crisis management through rational methods and technoscience are still vulnerable to the simplest forms of error and malfunction.

The narrative evolves around attempts to discover what killed an entire town (except a baby and an old man) after a satellite had landed nearby there. As in many other stories, such as the film *Outbreak* (starring Dustin Hoffman) and Robin Cook's *Outbreak* (on the basis of which the film *Virus* was made), the scientists are heroes fighting against a military-governmental inertia (and bad faith) to understand what is at stake. However, in Crichton's analysis, the heroes are extremely limited and only sheer luck provided a 'happy ending'. The emphasis on 'mutations' which subsequently exposes not only the failures of rational scientific reason, but also that of its technological applications, resonates throughout the literature on emergent pathogen virulence.

Whereas the AIDS pandemic generated a sense of socio-political urgency that entailed a radical multiplicity in popular cultural responses, *The Andromeda Strain* provided a 'generic' ethos for combining science and fiction with a degree of intensity that propelled both its popular appeal and its 'horrifying' potential. Of course, a single book can on its own hardly be held responsible for such a generic turn; however, the narrative it sets out resonates extremely well with the technocultural climate of our risk society. In itself, there is nothing new about a story of scientists trying to uncover the logic of an unknown, alien, deadly force – it has been part of most science fiction, even before cinema. Indeed, there is nothing particularly new either about civilian and military science joining forces to combat this kind of evil. What is new, however, is its attention to detail, providing an almost ethnographic account of laboratory practices and scientific reason. Indeed, as an ethnographic study of technoscience it was way ahead of its time. And what is most striking is that the horror is exactly in its detailed phenomenology of technological culture. Indeed, in the complexity of life itself technoscience reveals the fragile nature of human being. Hence the horror of *The Andromeda Strain* is that it is not resolved by the happy ending in which the scientists save their own skin and that of their world. The horror is just starting with the realization that this destruction might actually be happening as we speak. In this sense, the analysis breaks out of the enclosure of the literary genre and enters the very core of technoscience. The net result is an inevitable contamination of what was once deemed the rational world of technoscience as the flagship of the modernist project.

The Andromeda Strain not only opens a can of worms as far as technoscience is concerned but also extends this towards the apocalypse itself. In a way it precipitated a much needed alliance between science and technology studies and a sociology of the risk society; however it added a third dimension: that of a semiotics of apocalypse culture. Here, the link with AIDS has proven to be vital. The vectors of AIDS (at least the western variant) are often associated with 'evil' – sexual promiscuity (in particular when related to sodomy) and intravenous drug use. The very notion of 'unsafe sex' suggests strong connections with concepts of individualized risk and blame. Moreover, the way in which AIDS has become a multibillion dollar industry for test kits and medicine also points towards a fundamental corruption of 'the good' of medicine and public health. Indeed, the history of AIDS makes us aware of the radical evil, not only in the disease itself but also in its management. This radical evil is like a parasite engendering the exchange flows between risks and opportunities. AIDS has proven to be an uncanny ally in many strange alliances and actor networks in which human suffering is easily co-opted by economic, political, military and social opportunism (Oppenheimer, 1997). Indeed, the AIDS phenomenon also consists of an epidemic of 'signification' as Paula Treichler (1988) once called it. This epidemic affects mass media, popular literature, cinema, politics and last but not least the flourishing AIDS scholarship in academia – both in the biomedical and in the social sciences and humanities. AIDS has become a career for many researchers, scholars and critics.

The epidemic of signification mirrors the unintelligibility of the disease itself (Ironstone-Catterall, 2000). This unintelligibility is not just a consequence of the inability of virologists to provide concluding evidence about the way in which, for example, HIV infections develop into AIDS, or the way in which HIV seemingly evolves and mutates even within a single human body (Wills, 1996). This unintelligibility is more fundamental than that. It is attached to the logical impossibility to render an account of the motivation (used here in the widest sense of the word) of this specific form of pathogen virulence. By its very nature, every empirical science is limited by what it has cultivated as 'knowledge. Hence every encounter with an unknown force can provide an intelligible solution only if – in one way or another – one is able to render an account of the unknown in terms of what is already known. The already-known, however, is never of the same order as 'the objectification' which it produces; instead, the already-known is the symbolic order which provides us with 'sense'. In rendering an account of the world it encounters, it objectifies this world. The subsequent confusion between the objectification and the object, between the present-to-hand and 'the present', is what has allowed modern science to colonize 'reality'. That this colonization is actually derived from a confusion (between the empirical and the transcendental) needs to be forgotten.

Disease is unintelligible because it does not respond in the same terms as those of its objectification. In fact, it may not respond at all. No HIV has thus far been extracted from a human body, only the presence of antibodies have been visualized in laboratories. Signification, an endowment with signs, is nothing more than a displacement of 'the present' (which is itself always-already deferred) to the present-to-hand, that is, a discursive manipulation. At the core of this signification, there is a transgression: the transgression of the boundary between the real and the symbolic. We are condemned to such transgressions because our existence demands 'sense' to be made. However, at the same time, it becomes obvious that HIV, for example, is not confined to the laboratories of biomedicine, but has spread throughout the world; it has infected all cultural categories, and its infectious logic continues to wreak havoc. Like *The Andromeda Strain*, which has broken out of the literary genre, the assemblage of the book, AIDS too has invaded the immunity systems of our technological culture. AIDS is a disease, an industry, a career, a universal signifier that invites us to share a concern over our being in the world, and even invites us to condemn the evil that has brought it into our world. This evil maybe that of ecological destruction, biological warfare, pharmaceutical industries, fundamentalist zealotry, 'sexual promiscuity', moral indifference, carelessness or simply ignorance of the immense consequences. However, such condemnations always repeat the radical evil that has introduced them. That is to say, it perpetuates the chain of signification which – by disseminating blame, doubt, guilt and suspicion – ultimately brings down the very symbolic order through which sense is made of this unintelligibility. For example, in describing the social distributions of the consequence of AIDS on a global scale, many commentators pointed towards the inequalities of risk-positions, as well as to the way in which AIDS has become a vehicle for the dissemination of stereotypes (Patton, 1990; Lupton, 1994; Doyal et al., 1994; Waldby, 1996). Following a Foucauldean critique (often with critical annotations), these authors, rightly, point out that the disease has become an instrument of governmentality, discipline and regulation. In the margins of this critique, however, operates an often not very explicit suggestion that such appropriations are 'wrong' or 'dangerous'. However, at the very same time as such analyses are offered, we may also ask what motivates this obsession with regulation and control? Or, if the Foucauldean line is broken with, what motivates this obsession with struggle and resistance? What is being revealed in such critiques is the repetition of the radical evil that they seek to uncover. By offering an analysis of governmentality or resistance, is the academic not extending the epidemic of signification and in its wake, the self-destructing cynicism that motivates it? Of course, the same logic applies to this writing. This writing too becomes part of the epidemic of signification and its failed attempts to subdue the unintelligibility of pathogen virulence. It too is contaminated by the radical evil of risk opportunism. Self-valorization offers no escape from this infectious logic of abject virulence.

There is, however, something to be gained from such an analysis: the realization that we should always proceed with caution. Claiming to speak on behalf of the 'real' victims may seem like a noble deed; once this is a task that the speaker has claimed for him/herself it no longer is, especially when it is done 'to make a point'. *The Andromeda Strain* teaches us something not only about the fallibility of reason, but also of its political-institutional incorporation. It teaches us that virulence cannot be contained by knowledge alone; it teaches us to face the unknown, to acknowledge its unintelligibility, and forces us to seek means to converse with it, to respect its intelligence, and to avoid protecting oneself with a veil of cynical self-valorization.

Applying this to research involving emergent viruses it becomes obvious that we should acknowledge that there is evil. Apart from the possible lethal consequences of such pathogen virulence, the evil also resides in the institutions of management and valorization, including those of the biomedical and social sciences. Emergent viruses are a market, and a career, in more ways than one, as for example its popularity in recent films and novels shows. However, this does not mean that no good can come out of these attempts to make sense, to signify and to converse. Perhaps the ultimate truth of radical evil is that it can also be turned against itself, and thus can be made to destroy itself.

Enrolling virulence

Making sense of emergent pathogen virulence is a completely and thoroughly 'cultural' affair, even if performed almost exclusively within the technoscientific domains of virology (Cann, 1997; Christie, 1987; Dimmock *et al.*, 1990; Evans, 1982; Grist *et al.*, 1979), bacteriology (Salyers and Whitt, 1994) and epidemiology (Anderson, 1983; Bouter and van Dongen, 1991; Evans, 1982; Lilienfeld and Stolley, 1994). Moreover, technoscience cannot be constrained within laboratories, even if safely sealed off in biohazard phase-4 labs. It spreads into governmentality, the mediascape, commerce and, of course, affects us all, macrophysically as well as microphysically. This doubling of affect corresponds with Foucault's (1979) distinction between biopolitics and body politics.

In Chapter 3, we discussed at length the approach developed by Bruno Latour and his colleagues, which evolved around ethnographic studies of science and technology. Through these studies they developed a more general theory of 'actor networks' in which the emphasis lies on 'practices' of 'producing facts, effects and affects' that are simultaneously also practices of sense-making and 'ordering'. The great achievement in these studies is the enhanced ability to cut through the false and arbitrary distinction between reality and representation. Both are 'produced' as effects of technoscience and their ethnographic embedding in our being-in-the-world.

Moreover, these studies highlight the importance of the self-organization of actor networks as ensembles of relationships and connections between humans, technologies and gods, that are established forms of 'enframing' knowledge and experiences, and thus 'fixate' and 'fix' particular articulations as 'matters of fact'. The aforementioned 'AIDS establishment' can be seen as a particularly effective actor network; although it must be said this effectivity is more political and strategic than purely technoscientific, that is to say, although the network is very powerful, this is in spite of the lack of 'fixed knowledge' about the syndrome.

In the wider context of emergent pathogen virulence we do not find a single actor network that is as powerful as the one established around AIDS. This has nothing to do with the state of scientific knowledge itself; there is no more knowledge of HIV than there is of many other pathogens. It is the enrolment of public concern that has made much of the difference. People are more afraid of AIDS than of influenza or malaria, although the latter have claimed many more lives.

However, public anxiety on its own cannot explain the relative strength of an actor network. We need to focus more specifically on the mechanisms of enrolment and formation – who are part of the ensemble and who are not? In the AIDS network, HIV is very much part of the ensemble; the virus has been enrolled from the start to guide particular research and direct particular questions. The virological input into AIDS research has been phenomenal. One effect of this has been the shift away from a problematic derived from immunology (which translated into the counting of T-cells) to one derived from virology (the ascertaining of 'viral load') in clinical AIDS research. If we further take into account the enrolment of governments, media and commerce, then the relative strength of the AIDS network becomes more apparent; AIDS, and in particular its links with HIV and sexual transmission, have mobilized a concern from a range of 'actors' and 'technologies' that stretches beyond any other disease or epidemic. With so much at stake, it becomes obvious why there is so much 'fixation'.

HIV has been enrolled both in the biopolitics of population management and the body politics of self-discipline. The virus has been involved in the management of sexuality – the promotion of 'safe sex', curbing sexual promiscuity and generating a widespread anxiety about sex.[5] Moreover, the virus has also been enrolled in the management of the self and the body. Bodily excess, bodily fluids and bodily health are all regulated through imagined relationships to the microbiotic world. Hence, we see that with the coming of AIDS, new energy was invested into the maintenance of body boundaries and the intensification of an integral sense of self that coincided with the physical appearance of the body as 'individual'. This is highly reminiscent of the nineteenth century, where 'germ theory' prompted the state, often via middle-class 'social workers', to discipline the masses into a greater concern for hygiene and personal care (De Regt, 1984; Donzelot, 1977; Roderick, 1997). Despite the discursive embedding of disease in 'pathology', infectious disease has proven to be highly functional in the ordering practices of the state. Ordering populations into greater care for themselves, as well as 'the nation' would never have been as effective without the enrolment of pathogen virulence (Latour, 1988).

However, with the generation of a discourse of 'emergent viruses', whose existence is solely dependent on the cumulative effect of a greater number of 'unknown' and 'returning' infectious diseases, the functional enrolment of pathogen virulence might very well have come to an end. Whereas on its own HIV/AIDS might have prompted new codes for sexual interaction and the management of bodily fluids; its co-articulation with – for example – Ebola, Hanta, Lassa, MRSA and TSEs, creates a rather different balance between biopolitics and body politics. Here, we cannot escape the generic 'fixation' on the unmanageability of the microbiotic and within its wake, its undeniable associations with the general transformation of the global ecology. To a large degree, nation-state governmentality is not equipped to deal with global pandemics anyway, but when they multiply and are associated with very local vectors such as mosquitoes, rats and mice, the concept of 'public health' risks becoming a farce. With the growing sense of impotence of governmentality also comes a growing awareness that transnational bodies, such as the WHO in Geneva, are equally ill equipped to deal with the problem. As Ryan (1996) notes, there is currently (in 1995 – the year of the Ebola outbreak in Kikwit) only one specialist associated with the WHO who is responsible for emergent viruses. Claims for better 'monitoring' and 'surveillance' that are often offered as 'solutions' for the management of health risks (Henderson, 1993) thus become hollow statements, whose effect is simply the symbolic containment of public anxiety.

The 'sign of the times' is that the Foucauldean model of governmentality becomes increasingly divorced from the actual operations of the actor networks. Moreover, there is a growing suspicion that these actor networks are less and less 'governed' by agendas set by humans; instead it seems that technologies and viruses increasingly have the initiative. In this context, it is perhaps useful to reflect on the process of infection. The violence of infection is the violation of the body boundary, whose integrity is an investment of the individual to claim a sense of autonomy. This violation also opens up the body for further scrutiny by specialists and experts; once infected and handed over to medicine, the body further surrenders its autonomy and becomes an indiscrete element in medical technoscience. The infected body becomes incorporated into an already existing network. This network 'operates' on the body as a rendering of accounts (diagnosis) and functional modification (therapy).

However, perhaps we should ask ourselves who is rendering an account here. Is it the medic who operates on the body, or is it a wider constellation of medical technoscience that renders the body 'intelligible'? As beings of the modern age, we are used to placing ourselves at the centre of the cosmos – as both origin and destiny of knowledge, morality and progress (Foucault, 1970). We are trained to dismiss claims that perhaps it is not the human but a machinic assemblage (Deleuze and Guattari, 1988) that engenders 'history', as 'reification' or 'technological determinism'. However, what evidence do we have that humans are really the ones in control? Such evidence always returns to the same form, that of motivation: we humans must be the centre of the cosmos because we want to be. But when a cell is infected by a virus, the virus appropriates the infected cell's DNA for its own reproduction; it thus incorporates the body's genetic information, as it becomes incorporated into the body. This symbiosis is the essence of any parasitic relationship. The virus thus 'reads' the DNA as information for its own reproduction as it cannot 'live' without it. Hence, the virus is also rendering an account and is also reading the body's genetic information, to enrol the body itself into a specific actor network.

In the analysis of emergent pathogen virulence, there are sufficient doubts about the autonomy of human motivation to suggest that technologies and viruses are simply enrolled. They, too, have motivations which, moreover, remain unintelligible to us. These doubts should lead us to be suspicious about dismissing the roles played by technologies and micro-organisms in the constitution of our world. Consequently, when rendering an account of infectious disease, or public health risks, it cannot be sufficient to point towards biopolitics and body politics as purely 'human' concerns. It is vital we keep an open mind towards sensibilities and sense-making practices that disrupt our will-to-know as the sole guide of our destiny.

Conclusion

At first sight, the newly emergent viruses, as well as the recurrent older ones, seem to provide excellent support to Ulrich Beck's risk society thesis. They are largely invisible (at least without the aid of highly elaborate and expensive electron microscopes), yet ubiquitous and terrifying, sometimes even lethal. Moreover, their emergence coincides with global political, economic, social and cultural developments that may signify an 'end' of industrial modernity and its socio-political anchorage in the nation-state. Above all, their emergence cannot be dissociated from a more generic global ecological crisis, as the hot zones of epidemic outbreaks are often the same marginal zones of industrial development and ecological exploitation. They coincide with climatological changes, pollution, mass poverty, the destruction of rainforests, large hydro-electric projects, new motorways, hospitals, social and political unrest, wars, famines and mass migration.

Moreover, it seems to provide further support to the thesis that the risk society inaugurates a turning in technological culture in which science and fiction, innuendo and matters of fact, evidence and speculation all become equivalent signifiers in a frantic global spectacle of sense-making (sometimes called 'postmodernism'). The failures of western institutions to control the rise of epidemics, and the public anxieties that are raised in the wake of techno-scientific catastrophes, further contribute to a sense of despair. The science-fiction and fiction-science writings on Ebola, for example, all breathe the same ethos of doom.

What we get in these science-fiction hybrids is a paradoxical apocalyptic utopia in which fear and anxiety are being set into work via ever more rational solutions to deal with the problems at hand.[6] Such paradoxes are resolved only by negation, that is, by ignoring that there is a paradox. Towards the later parts of her chapter on the end of antibiotics, Laurie Garrett (1994), for example, seems to have actively forgotten her initial mapping of the relationship between mutating microbes and antibiotics as one of shifting transgressions, in which the defeat of antibiotics was immanent due to the superior capacity of bacteria to overcome antibiotic treatments (it is part of the genus – habitus – of bacteria to do that). Instead she suggests that misuse of antibiotics is the main culprit, in particular in 'Third World' countries. If the latter were true, then 'solutions' are easily found: education, training, discipline, monitoring and control. However, if the former claim is true, then such strategies might only be delaying tactics – eventually, the microbes will catch up (Cannon, 1995; Garrett, 1994).

However, there is clearly no reason why we should not be concerned about the organization of public health in the face of the end of antibiotics. The delaying tactics now deployed, for example, in the treatment of malaria (the latest – most effective – treatment will be given only after all other ones have failed), will help to sustain some effectivity for a longer period of time than without such planning. There is also little to disagree with when considering the efforts of devoted virus-hunters, physicians and bacteriologists who focus on educating and training local physicians and medics in many of the poorest regions of the world to recognize particular infections and deploy the most effective treatment known in (western) medicine. However, when taking a slightly broader view, one clearly sees that such efforts work only in their particularity. So, for example, it is against the logic of capital to develop policies aimed to offset rising costs of medication, in particular those of antibiotics. Accordingly, the US Pharmaceutical Manufacturers Association has played a key role in blocking many of these initiatives in the name of 'free trade' (Garrett, 1994: 438; Mann et al., 1999).

If we take the full picture, that is, one including not only pathogens and medicine, but also the global economy, capitalist industry, the warfare complex and, last but not least, apocalypse culture, it becomes obvious that there are no easy solutions. Indeed, there is no reason to be optimistic. At the end of modernity, we now fully encounter the madness of reason as it has taken on 'lives' of its own. The madness, which is the radical evil that corrupts any semblance of 'good judgement' (Rogozinski, 1996), prevents us from 'doing the right thing'. Within the limits of reason alone, we are likely to become subjected to a microbiotic revenge that will – in the long run – decimate the human presence on this planet to such proportions that it is no longer able to threaten what Garrett (1994) describes as 'world out of balance'. However, what lies beyond reason, beyond our abstract machine of technoscience, could very well entail the saving power towards which our predicament is destined.

8 Cyberrisks

Telematic symbiosis and computer viruses

You know – technology wasn't invented by us humans. Rather the other way around. As anthropologists and biologists admit, even the simplest life forms, infusoria (tiny algae synthesized by light at the edges of tidepools a few million years ago) are already technical devices. Any material system is technological if it filters information useful to its survival, if it memorizes and processes that information and makes inferences based on the regulating effect of behaviours, that is, if it intervenes on and impacts its environment so as to assure its perpetuation at least.

(Lyotard, 1991: 12)

Viral epidemics present a case of abject virulence which has co-evolved with humans, animals and plants since the very beginning. Their existence as such is not specific to modernity's technological culture. However, we argued that 'emergent pathogen virulence' is a typical example of the 'turning' of technological culture, indeed perhaps even a messenger of the incumbent apocalypse. This is because the logic of emergence is intertwined with the limits of technological culture. As it expands into previously unknown 'junctural zones', especially tropical rainforests, this technological culture challenges and reveals nature's darkest secrets. Moreover, the continued exploitation of life, not just in terms of capitalism but in terms of the very underpinning of the technical systems that 'run' this economy, including its submolecular level, has created 'hot zones' of abject virulence, marked by high levels of pollution, genomic interaction and infection. However, all these instances of abject virulence could be summarized by the subtitle of Geoffrey Cannon's (1995) book – *Nature's Revenge*. It is based on the idea that 'man's' asserted dominion over nature, through the deployment of technoscience, is now ending. Emergent viruses are merely the payback of decades of unbridled and careless commercial exploitation of nature, especially the rainforests.

Nature has interesting ways of balancing itself. The rain forest has its own defences. The earth's immune system, so to speak, is seeing the presence of the human species and is starting to kick in. The earth is attempting to rid itself of an infection by the human parasite.

(Preston, 1995: 366–7)

This view, however, has its own limits. It assumes that nature itself is a coherent, homeostatic system (i.e. ecosystem) that, if not motivated by retribution, is motivated by a desire to restore its own order. This view contradicts the evolutionist perspective which understands nature as a process of continuous transformation. Moreover, it sets out a distinction between humans and nature, as if the former are separate from the latter and thereby fails to acknowledge the ambivalence of nature, as itself following the logic of a technical system, be it that of homeostasis or evolution. Moreover, this view is problematic because it assumes 'man' to be merely a catalyst in the unfolding of otherwise completely natural risks. It forecloses any understanding of risks as part of an assemblage of continuous flows of differentiation and modulation. Finally, such a view is deeply misanthropic. It rests on a Malthusian domesday scenario which sees overpopulation as the source of all evil.[1] The eugenic misanthropy is particularly directed at the Third World, where the problems are most severe, yet thereby negates the central role played by the affluent west, whose wealth is largely derived from worldwide exploitation. It thus endorses a highly questionable politics of 'blaming the victim'.

Instead of promoting a misanthropic notion of nature's revenge, this chapter seeks to widen the 'turning' of technological culture in the risk society by looking at risks that can never be idealized as 'natural'. The starting point is a problematization of 'nature' based on Macnaghten and Urry's (1998) work *Contested Natures*, which will be broadened into a problematization of 'the environment' as a whole. Appropriating the concept of autopoiesis, it will be made clear that a cybernetic conception of life as information processing will not only annihilate the unhelpful distinction between nature and culture, but also enable a conception of technological culture as entropic. Abject virulence is thus not a regulating device of systemic integrity, but a radically transformative and vital force in the reconstitution of life. The concept of 'cyberrisk' will be invoked to open up the scope of analysis to the material-informational technological processing of risks that comes before their discursive realization in signifying practices.

Exposing 'the' environment: a cybernetic turn

In his acclaimed *Rise of the Network Society*, Manuel Castells (1996) describes the contours of an informational landscape characterized by a series of relatively smooth, globally operational, flows (of data, people, money, goods, services, images). This landscape, or what he calls 'space of flows', is the field in which global capitalism (with its geopolitics of the new world order) is able to bolster new heights in the accumulation of wealth and power. The space of flows, however, is further marked by an increased separation from what he calls 'the space of places' – the lived geographies of everyday life. It is the growing gap between these two landscapes, the space of flows and the space of places, that is the definitive condition of inequality in the informational society (Castells, 1996: 428).

Irrespective of whether one agrees with Castells's vision of the future or not, his notion of an informational landscape is worthwhile pursuing. It highlights the problematic of conceptualizing 'the environment' in an age in which technoscience has irreversibly imposed its own specific logic onto everything else. In this chapter, I will argue that the environment is itself the product of an internalized differentiation of specific systems. The environment is a label that is assigned to that which remains; it is the by-product of the technological enframing of the world that has been called 'modernity'. In this sense, the term 'environment' shares the conceptual space reserved for 'waste', 'dirt', 'residue', 'excess' and 'risk'. It is thus intrinsically contingent upon those systems which order the world: technoscience, governance (including politics and law), media and commerce.

Throughout this book it has been argued that technology reveals and orders the world. It produces a specific relationship between the existential being (what Heidegger (1927/1986) calls *Dasein*) and its being-in-the-world. This 'in-the-world' could be conceptualized as basically an 'environment of being', a space in which being emerges. It has also been argued that technological culture orders and reveals a specific relationship between human being and 'his' environment. This relationship is principally enframed in terms of a challenging-revealing of the environment in support of 'man's' full self-realization as an autonomous being. That is, technology offers itself in service of 'man' and constitutes a relationship between 'man' and 'his' environment in terms of what it orders and reveals.

However, technological culture does more. It enables 'man' to encounter 'his' environment carelessly. That is to say, it engenders a 'forgetting' of concern for being outside the technical system which mediates the encounter. Technological culture generates enclosure through mediation by ordering an environment. This is clearly shown by Macnaghten and Urry's (1998) critical analysis of the way in which different conceptions of 'Nature' are actualized in social practices. The different conceptions of 'Nature' simultaneously posit 'it' as a singular, integral and self-evident entity. That is, whereas there have always been competing constructions of Nature, the incompatibilities have only relatively recently become subject to more overt political struggle. Indeed, even if the dominant conception of 'Nature' as an externalized environment still prevails, the contestations of such a conception have now permeated all aspects of technological culture.

The concept of 'virtual object' has been used to argue that the process of 'enpresenting' involves a complex chain of revelations and orderings that engender a multiplicity of actualizations. We distinguished between presentation, representation and re-presentation to show that technoscience is merely one way of 'making present' and albeit the dominant one in technological culture, it cannot foreclose alternatives. Technoscience is mostly concerned with representation and as a result operates on virtual objects *as if* they are real. Until recently, scientific realism was largely hegemonic and hardly contested, at least outside the humanities and social sciences. Environmental risks are simply conceived as hazards with probability calculations attached to them (Cutter, 1993; Fox, 1998: 671–3).

This, however, can no longer be taken for granted. Whereas there have always been sound philosophical arguments against the naïve realism of risk assessments, this chapter argues that there is now an overwhelming historical-realist case against it as well. This argument, perhaps originally posed by Lyotard (1979) in his playful, perhaps even misleading, concept of postmodernism, was further developed by theorists such as Baudrillard, Jameson and Vattimo to stress that the very same technologies of representation which inaugurated technoscience as the superior mode of revealing, have now turned against the very foundations of this power. Especially, with the advent of digital information and communication technologies any stable and fixed notion of 'reality' has been undermined. The Gulf War, for example, gave us images of 'smart' bombing that may actually have been produced by a simulator. Images of Palestinians dancing in the street after the news of the bombing of the World Trade Center on 11 September may have been more than ten years old. We live in a hyperreality. Our environment is now so heavily mediated that we cannot even trust our own empirical senses to give us direct access to the real.

This 'real' used to be the environment of our being in the world. However, the environment of ICTs is always-already subject to continuous virtualization. Digitalization, especially, is a force that shifts power away from atoms to bits (Negroponte, 1995). For example, it is telling that genetics is modelled on information science (in particular the work of Shannon and Weaver, 1949), not the other way round; this dates back to the early 1950s. Our very concept of the logic of life is modelled on technology. This is not new. The Greek word *organon*, which serves as the basis for organism as well as organization, originally referred to machinery.

For the world of ICTs, the environment is always a virtuality – which can be realized only through deferral. The space of flows of the informational landscape discloses 'an' environment that is endlessly deferred in its enframed 'out-there-ness'. The contingent singularity of this endless deferral, or differance, is nothing more than the actualization of a particular technology, or medium. ICTs have also enabled a new kind of sensibility within technological culture. This sensibility is marked by a rapid acceleration of information flows: the future becomes increasingly 'present' as an extended presence (Nowotny, 1994). This has created a possibly new kind of episteme which could be called 'hyperthought'.

The difference between modern thought and hyperthought is that whereas in the first episteme environment remained contingent upon integral or molar media-forms (science, governance, media, commerce), in hyperthought environment is enacted by a perpetual differential or modular assemblage of media-forms (science-becoming-governance-becoming-media-becoming-commerce).[2] Indeed, it is the modulation of the different flows that spatialize the informational landscape into particular media assemblages that characterizes this new hyper-contingent sense of 'environment' (Clark, 1997b: 58).

The deadlock between reality and representation, referent and sign, objectivity and subjectivity that constitutes the episteme of modern thought is suddenly broken with the advent of new information and communication technologies. Indeed, the world has changed in many ways. Of course, we still have capitalism, we still have nation-states, parliamentary democracy, disciplinary power, racism and patriarchy, but we must seriously question whether any of these labels still 'signify' the same 'phenomena' (and inevitably, whether the very labels 'signification' and 'phenomena' are what they used to be).

We should not be misled in thinking that everything should be different in order to find a rupture or discontinuity. That is a typically modernist presupposition ('a rupture is when everything changes'). Perhaps the difficulty we face now is that change is not as sweeping, clear-cut, totalitarian and all-pervasive as 'the Great Transformation' (Industrial plus French Revolutions). It seems as if we are in the midst of a perpetual rupturing as new worlds ('the' informational society being one of them) are hailed and forgotten by the flick of a page (or screen). Modernity is valorized by 'the' new but in a world that has lost the unity of perspective, 'the' loses its special particular-universalist character. Speed is so intense that the very possibility of sustained newness is vanishing rapidly.

When considering an environment such as that of new information and communication technologies, we cannot but account for its technological actualization. We may project such an environment to be everything 'outside' of the technological enframing of information processing and digital communication, it could even encompass the entire planetary physical world, including the media worlds of books, cinema, television and radio; however, this is possible only if we also acknowledge that which is 'inside', or more accurately, that which produces the difference between inside and outside, i.e. *autopoiesis*.

Social theoretical conceptions of autopoiesis often operate within a scientific paradigm founded by system theorists such as Karl Deutsch and Norbert Wiener. The latter devised a term for it: cybernetics. It is this paradigm that may be seen as fundamental to the development of artificial intelligence (AI) in the 1970s and the subsequent information-processing and telecommunications revolutions of the 1980s. Cybernetics concerns the organization of the processing flow of information and the ability not only to reproduce itself, but also to adapt to and modify its environment in order to secure sustainability (Tomas, 1995: 23). Quite clearly, autopoiesis is a fundamental process in cybernetics – it is the becoming-self-reliance of an information-processing device. However, cybernetics also highlights something that Luhmann and Teubner seem to ignore: namely *symbiosis*. For the theorists of autopoiesis, life is a process of chance – a coincidence established on the basis of the unlikely autopoietic take-off, when an information-processing device becomes an organism. However, studying viruses for example, it is quite obvious that life is not exclusively limited to autopoietic forms. The life of viruses is dependent upon its capacity to live alongside its host, that is, to establish a symbiotic relationship (Ryan, 1996: 279–80).

It is symbiosis that allows systems to induce each other, to transform their information-processing flows. This is the core of what we are experiencing as the telematic revolution. Telematics refers to the interconnection between systems of telecommunications and information processing, allowing for instantaneous, worldwide dissemination of data. It operates on the basis of a very simple principle of digitalization. Whatever can be transformed and reproduced in digital form can be processed via telematics – text, sound, image, even neural and tactile-sensory experiences. It is thus not so much the arrival of new media, but the *assemblage* of media technologies, that marks the telematic revolution. The assemblage combines simplicity with speed. It allowed for new powerful intersystemic symbioses as media technologies have started to pervade the military-industrial complex, large corporations, the state apparatuses, the culture industries and increasingly our private homes. Satellite, telephone, cable, computer, television operate together; the information flow allows for the transcoding of all kinds of other flows: finance, signs, goods, services, even people (Castells, 1996; Hillis, 1996).

The speed with which the telematic symbiosis has established itself is phenomenal; and there is no turning back. The transcoding is infinite as nanotechnologies rewrite DNA – 'nature' becomes just another series of ones and zeros. 'The' environment, which was once discursively protected by a clear-cut and intransgressible distinction between nature and culture, is now up for grabs – open to any form of decoding and transcoding. Consequently, we would be quite mistaken to separate an environmentalist discourse about the natural world from discourses of telematic symbiosis.

In the midst of the speed and irreversibility, 'man' – who already lost a centred consciousness in twentieth-century philosophy and psychoanalysis – is now also rapidly losing the plot in managing information flows. Reading has become browsing, reflecting now takes place by the move of a cursor on a hyperlink and the mere click of a mouse. Telematic symbiosis has surpassed our ability to master and control it and subject it to our programming of the future; instead our futures are being programmed as we speak. This sense of loss manifests itself in the uncanny but perpetual return of the same debate over the merits and evils of cybertechnologies. On the one hand, the technophiles and technocrats cannot stop exaggerating on all the goods of the virtual: we are no longer hampered by the inertia of the flesh, construct our identities as we please, are free to engage in whatever we like and become participants in a more open and equal form of civic society (Kelly, 1997; Negroponte, 1995; Rheingold, 1991, 1994). On the other hand, the technophobes and neo-luddites do not stop complaining about the dumbing and numbing effects of telematics. Telematics as nothing but an extension of capitalist exploitation, politics of propaganda and mystification of western imperialism, turns us into apathetic, lethargic, brainwashed consumers of infotainment (Dovey, 1996; Robins, 1996). Both extremes show a remarkable inability to sensitize the inherent ambivalence of the telematic symbiosis and operate as if in a permanent neurotic narcosis. Both extremes are still trapped in a technological culture, governed by instrumental fidelity and humanist (or its mirror image: misanthropic) delusions (Ansell Pearson, 1997). They are not different from the arguments for and against technological innovations in the nineteenth century. As we have learned, technoscience continued to march on and revolutionize the world.

However, there are grounds for a (tempered) optimism as well. Unlike in the nineteenth century, technoscience is no longer unified by a single episteme (spearheaded by positivism). Cracks and fissures have appeared and have induced more critical sensibilities. Risk has now revealed limits to technological culture that cannot be brushed aside by a mere reference to the inevitability of progress. It is here that the connection between telecommunication and data-processing has been most effective. Whilst being generated by a desire for speed and overcoming obstacles of time and space, telematics have also engendered new forms of alienation. Whilst initially such alienation could be seen as a primary force for identity-politics and projects of the self (Giddens, 1991), an increasing number of people have given up attempts to unify themselves or the relationship between their bodies and subjectivities. This loosening of univalence between the body and subjectivity enables human beings to live their lives as decentred creatures. There is no need to read Derrida or Freud to understand the decentring of the subject; it has become common sense.

It can be argued that the decentring of the subject finds its actualization not in philosophy but in telematics; it is here that the subject can finally come to practical terms with the externalization of her/his nervous system which – according to McLuhan (1964) – took place with the arrival of television. A direct consequence of this, it can now be argued, is that the relationship between human beings and their environment is changing. The notion of 'man' as centre of the cosmos now looks as ridiculous as the pre-Copernican claim that the earth was the flat centre of the universe. It is less clear, however, what or who is to take 'his' place.

The space of flows, or cyberspace, is not only a particular type of environment, but also one which problematizes the 'naturalistic' notion of 'the' environment and the implicit centredness of human consciousness therein. Environments are produced remainders of systemic self-referential differentiation (autopoiesis) and intersystemic symbiosis. They are formed through a process of territorialization, in which an assemblage creates a milieu, by recoding elements from other assemblages, that is, by deterritorializing other environments (Deleuze and Guattari, 1988).

The telematic revolution of the 1980s has exposed the inherent flaws in conceptualizing 'the' environment as an ecological 'natural' system (also see Clark, 1997a). It is the transcoding of flows between the systems of science, governance, commerce, media, law and the military that marks the current climate of conceptualizing environmental risk. Whereas these systems may continue to operate on the basis of their own distinctive self-referential technologies and their inherent logic of, for example, secrecy, expertise commodification or mediation, it has become obvious that the real power is exercised from the assemblage that emerges in-between/beyond them – in the symbiosis of technologies of visualization, signification and valorization. Only when these forces are adding up, will a virtuality become 'realized'. Such a symbiosis is thus a closure of entropy, a gathering and ordering of multiplicities into a singular assemblage. In technological culture, this singularity was achieved through the domination of technoscientific codes of representation. Technoscience revealed and ordered the real. In the risk society, this old symbiosis of representation is eroding and has started to turn against itself. The newly emergent telematic symbiosis thus entails a deterritorialization of the previously closed milieu of technological culture.

A key feature of the assemblage is that both system and environment constitute each other interactively. Hence, it follows that if we argue that telematic symbiosis engenders a new multiplicity of (contested) environments, then this itself impacts upon the 'media systems' which have engendered the emergent symbiosis in the first place. Indeed, as with 'the' environment, we are witnessing some radical changes in the systemic organization of the 'mediascape'. As mentioned in the previous section, we are witnessing a rapid hybridization of media-forms which have sent panic waves of ambivalence throughout modern western societies, undermining the fixed boundaries and classification systems through which discipline and capital could organize and valorize a particular social, cultural, political and economic order.

What characterizes this mediascape is a growing uncertainty over the boundaries between on the one hand the real and the possible, and on the other hand the virtual and the actual (Shields, 1999).[3] The telematic symbiosis of the mediascape has affected all 'domains' of disciplinary society: science, governance, law, the military, commerce, and on all levels from corporations and institutions to individual bodies and even genes. The blurring of the boundaries between the real and the possible has generated a crisis over realization. Increasingly, government agencies and corporate bodies are operating on the basis of complete uncertainty whether something can be realized or not. As the intervals of action are shrinking, time is vanishing. The speed of anticipation starts to overtake that of actualization. Technoscience, with its carefully cultivated and protected enclosures of expertise, is simply not up to the task. Governments cannot wait for laboratory experiments to confirm with 'scientific certainty' that something is a 'real' hazard or not. As definitions of what is hazardous are deeply political (Fox, 1998), a loss of 'technoscientific sanctimony' may actually be advantageous. It may open up debates about the perspective-bound relativity of risks and hazards. However, this optimism could also be premature, as new forms of subpolitics – judicial, military, commercial, media, administrative – are already waiting and eager to fill the void.

Moreover, the blurring between the boundaries of the actual and the virtual has engendered a crisis of actualization, in which the distinctions between fact and fiction, or the referent and the signifier, have become increasingly void of any differentiating power. The speed with which these boundaries are being blurred by the telematic symbiosis has caused a number of serious breakdowns in social organization. One only needs to look at politics or stock markets to understand the volatility of signification, as Peter Bernstein testifies:

> Capital Markets have always been volatile, because they trade in nothing more than bets on the future, which is full of surprises. Buying shares of stock, which carry no maturity date, is a risky business. The only way investors can liquidate their equity positions is by selling their shares to one another: everyone is at the mercy of everyone else's expectations and buying power . . . Such an environment provides a perfect setting for non-rational behaviour: uncertainty is scary.
>
> (Bernstein, 1996: 300)

In analyses of media coverage of 'environmental' issues, what generates most attention is the 'representational' aspect of coverage, that is, the signs that are brought into presence in the place of whatever they are made to refer to (Allan *et al.*, 2000b; Anderson, 1997; Chapman *et al.*, 1997; Hannigan, 1995; Hansen, 1993; Mormont and Dasnoy, 1995). Coverage in this sense refers to a process of displacement: real environmental risks are replaced with signs; the signs are stand-ins for 'the real' (Eco, 1977; Pêcheux, 1982). 'Critique' in this sense often refers to the adequacy of this 'standing in' – do the created signs actually correspond to that which they have replaced? Quite obviously, the crisis of actualization, which is inherent in the telematic symbiosis that is redesigning many mediascapes, obliterates any firm grounding of distinctions between the real and the representational, since both can only be expressions of a limit condition of the virtual.

Fortunately, most critical analyses go beyond this and ask a much more productive question of what such media coverage *does*. Here we enter into a second axis of understanding representation, this time not as a process of replacement, but one of displacement. Representation is then not so much a confinement of the real in signification, but a specific realization of possibilities. Mediascapes are articulations of reality effects. Hence, the critique shifts from a question of correspondence to one of performativity – what is the coverage actually achieving? In media studies, answers are most often located in terms of 'covering-up'. Here, the media are seen as ideological state apparatuses that mystify environmental hazards and naturalize them (Coleman, 1995; Hannigan, 1995).

Although there are clear reasons to argue that environmental hazards are often covered-up in mediascapes (see for example Allan, 2000), as information about risks is often controlled by actors who may have a lot to lose, we do not need to place our sole faith in conspiracy theories. The point to make is that 'coverage' or representation does more – it is not only an act of concealment, but also one of unconcealment (in terms of sense-making). This requires a more productive (performative) understanding of mediascapes as particular technologies. As we have seen in previous chapters, media representation operates on the basis of technologies of signification, thereby allowing other systems (including what we may call 'the public') to articulate and relate to perceived risks in particular ways – and thereby engaging in what may be called the (sub)politics of the risk society (Beck, 1996, 1997).

Virtual risks and telematic symbiosis

The question to be asked here is in what way can Beck's risk society thesis speak to an 'environment' that has already been completely incorporated by technologies of mediation? In previous chapters, we have argued that it is in the very ambivalent role of technology/mediation – as both a making-sense (unconcealment) and a re-presentation (concealment) of risk – that the risk society reinstates the paradox of modern life as an order of disorganization. The power of Ulrich Beck's work on risk resides mostly in its visionary ability to see that we are witnessing a cultural-transformation of risk perception. This cultural transformation is predominantly, but not exclusively (see Lash and Urry, 1994; Welsh, 2000), organized around a particular rupture – the legitimation crisis of technoscience. The risk society calls upon the very same technoscience to both reveal and conceal the danger; the paradoxical consequence being that of technoscientific delegitimation and political indecisiveness – and a further eradication of the foundations of modern society.

It is now common sense to associate technoscience with risk. Associated with both causes of and solutions to risk, science and technology have played an intricate part in the construction of economic, social, political and cultural conditions under which risks operate in modern society. Technoscience primarily operates through technologies of visualization – thereby generating particular perspectives and insights and constituting the condition for (mis)recognizing unspecified possibilities as 'risks'. Governance and mediascapes tap into that visualization with their respective technologies of signification: governance by articulating a distinction between 'significance' and 'insignificance' (hence difference and indifference), mediascapes by articulating sense and cultivating sensibility, bearing witness to the event of visualization and (governmental) signification as it unfolds.[4] It is often suggested that media invoke technoscientific sources to grant legitimacy to the coverage (Coleman, 1995: 68). The acquisition of such a legitimacy, however, is not the accomplishment of either science or mediascape, but of the assemblage between technologies of visualization and signification – the first by giving insight, the second by attributing meaning.

To signify is to attribute meaning. The attribution of meaning is a form of encoding, of 'recovering' that which has been revealed. The stabilization of this recovery is the work of *logos*. *Logos* can be made to stand in for a range of other terms: language, law, logic, even reason or discourse. It always refers back to 'coverage' or 'concealment'. When something is put under the heading of *logos*, it is given a name, it becomes an attribute of the name, it becomes covered by the name. As Foucault (1970) pointed out, modern thought went even a bit further than merely 'naming' and 'classifying' and started to replace the name (language) with 'logic' (both in terms of causality and of law). This can be traced in today's usage of the term ecology, which often implicitly refers to a carefully balanced 'natural' system in which the laws of nature work in a perfect symbiosis to establish harmony and homeostasis.

Nowadays it is impossible to operate under the heading of ecology without submitting oneself to the technologies of visualization that characterize much of modern technoscience. Granting 'insight' as the basis of knowledge is the central focus. Even if there is a considerable and critical residue of knowledge outside of the technoscientific establishment (cf. Welsh, 2000), one cannot underestimate the powerful impact of the technoscientific overcoding – or overexposure, as Clark (1997a) calls it – of such knowledge. This is because technoscience has been very effective in symbiotically appropriating the force of technologies of signification. Technoscience has successfully induced both governance and the mediascape (as well as commerce, but I will not focus on that here) to be able to intervene in the differentiation between significance and insignificance. This kind of 'subpolitics' does not engage with the same forms of accountability that characterize the political domain of representational democracy. What is nowadays called the public understanding of science can be seen as a programme of inducing the mediascape in order to more effectively operationalize the technologies of visualization (Irwin and Wynne, 1996; Rose, 2000).

What happens to environmental hazards when they are overcoded by technoscience is that they are being transformed into objects of visual desire. The desire to make visible the implicit causal link between substance A, malign effect B, and possible effective treatment C (although the latter is much less clearly in the picture, it remains a promise) is a major motivation of contemporary scientific 'discovery' (Welsh, 2000).[5] What happens as a result of this visualization, however, is the generation of more technoscientifically accurate 'knowledge' about hazards, which, via the enrolment of governance and mediascapes, are transformed into (sub)political issues for decision-making via signification and become risks (Beck, 2000; Cutter, 1993; Fox, 1998; Luhmann, 1995).

With 'the' environment already transcoded by the telematic symbiosis, and signification being induced by technoscientific desire, it is quite difficult to see any relevance in asking the question whether such environmental risks are real or constructions (Beck, 2000; Van Loon, 2000a). They are always 'virtual'. What matters is the way in which these risks are being transformed: environmental risk becoming health risk becoming economic risk becoming political risk becoming representational risk (for example, the BSE crisis in the UK and Europe). In a world of perfectly autopoietic systems (such as the disciplinary society) such risks would remain enclosed within one (sub)system only. With telematic symbiosis, however, they can move relatively smoothly across a range of intersystemic boundaries and at high speed. The result is a hyper-intensified multiplication of risks.

Hence, environmental risks are no longer contained by any single particular system, be it that of science, governance or commerce. Telematic symbiosis has allowed risks to flow across systemic boundaries, inducing a range of environments and further contaminating and corrupting systemic integrity. Indeed, far from being an event of technocratic supremacy, the assemblage that is currently being formed between technoscience, governance, media and commerce bears all the signs of a highly vulnerable and fragile stature.

Hence, it seems perfectly logical to extend Ulrich Beck's analysis of risk society to the domain of electronic communications. Just as there have been applications to nuclear, genetic and reproductive technologies, so can the risk society thesis be applied to the world of information and communication technologies. Particular examples of such risks would be moral panics, such as outbreaks of 'irrational fear' after particular broadcasts, e.g. XTC-deaths, BSE, youth cultures (e.g. Cohen, 1987; Hall *et al.*, 1978; Thompson, 1998), information overload and noise (e.g. middle management stress), cyberaddictions and cybercults even resulting in mass suicide (HeavensGate), hacking and cyberfraud (e.g. with credit card numbers). Finally, of course, the hallmark of cyberrisks is the computer virus, which represents, as Nigel Clark (1997a: 79) puts it 'a new generation of digital demons to suddenly render our entire datasphere into terminal gibberish'. In order to understand the material-semiotic performativity of the mediation of environmental risk (that is, the question of how environmental risks operate when overcoded by a telematic symbiosis), we need to engage with the concept of cyberrisks.

Cyberrisks

> The simulation of nature is reaching this level of 'sophistication', raising the possibility of simulacra 'running wild', interacting, and even generating their own forms and functions . . . 'irruptions' of simulated nature now appear to offer both the scientist and the general consumer of images the enchantment of a 'living' world which is constantly, rapidly, endlessly generating novelty.
>
> (Clark, 1997a: 86)

The risk society thesis works on the basis of the assumption that the technological culture inhabited by humans can no longer perform functional closures. That is, the self-evidentness of the ordering-revealing enabled by technoscience can no longer be taken for granted. As a result the relationship between human being and environment is no longer unproblematically mediated by technology; instrumental fidelity breaks down (see Chapter 10). This breakdown is catalysed by the digitalization of environments. In the world of telematics, the relationship between human being and environment is inevitably ambiguous. The environment is already multiplied by the telematic assemblage and this inhibits a taking for granted of its unified nature.

Whereas in a pre-digitalized technological culture, the virtuality of risks still needed to be argued, usually against realism, for telematics the virtual has already taken over. As telematics fundamentally disrupts the categories of modern thought, risks also become inherently destabilized by the sheer force of virtualization. A simple example would be the 'I Love You' computer virus which, apart from directly jamming network systems, also caused immense indirect problems as system operators and webspinners across the world had to generate warning messages and upgrade virus detection systems. This is an example of a risk that is able to reproduce itself not simply through failures of regulation and security, but exactly because of the risk-aversion ethos already put in place by the systems themselves. Once risks are able to work through successful risk-management systems and use them to proliferate themselves, that is, once they are not only able to act but also to anticipate and learn, we could speak of cyberrisks.

The notion of virtual risk has been used to implode the realist/constructionist distinction between the real and the representational (Van Loon, 2000a). Beck's notion of risk involves a complex alignment between very particular and very exclusive modes of signification (most notably those of 'applied' and 'pure' science), with financial, economic, judicial, political and administrative institutional systems of representation. Beck's rather exclusive focus on 'institutional' risk perceptions must be supplemented with the popular and allegorically organized moral and practical *sense* that risks make in everyday life settings. Risks are realized in the experience and appreciation of a variety of particular discursive and figurative alignments. This necessitates that risks must have an *appeal* to an audience to be made sense of as risk. Risks thus involve an interpellation: a calling upon potential receivers to respond (without ever, of course, assuming only one possible response). In an age where television is capable of instantaneous worldwide dissemination of information to an almost limitless mass of receivers, the appeal of risk is formed in a cultural matrix which exceeds that of technoscience and institutional politics: it has the capacity to 'directly' intervene in the social density of everyday life. Not all 'risks' need to be mediated by either expert knowledge or locally embedded moral consciousness to operate as risks.

Although the essence of risk always entails both an understanding of probability and intensity of harm in relation to a specific decision-making capacity, neither probability, nor intensity, nor harm, nor decision-making can be exhausted by their semantic operators. This is because semantics exclusively concern *logos*, that is, the reasoned logic of signification. Supplementing this *logos* is the possible availability of *nomos* – the practice of distributing intensities across a field of forces and intensities that is not its own (Deleuze and Guattari, 1988: 372). *Nomos* is distinct from *logos* in that whereas the latter always searches from the principle of reduction, of a reproduction of origin, *nomos* is a flow of deviations and irreducibility. *Nomos* is the logic of escape without disappearance or sublation. It cannot be controlled by the sign. This *nomos*, or what some poststructuralists have called 'decentred subjects', generates the non-isomorphic deterritorialized supplement of virtual risks – their transformation into a material force that is no longer exhausted by its signifying semiotics (Haraway, 1988). In short, cyberrisks – virtual risks of the telematic society – are self-replicating actors that are not mediated by consciousness.

Virtual risks are the particular expressions of experiencing risk perception. These expressions concern the discursively mutated 'sense' that risks and hazards make in the everyday lives of human beings, gods and technologies. Virtual risks are inscribed into the Abstract Machine of modern society where the signs in which they exist are being connected to 'a whole politics of the social field' (Deleuze and Guattari, 1988: 7). As a special case of virtual risks, cyberrisks are multiplicities from the start;[6] they exist primarily as 'determinations' that cannot be reduced to any code existing beyond itself. The code itself is the pathological agent. There is no 'pure form' of cyberrisks; they are 'only' mutations and deviations.

Cyberrisks have no 'history'. As they do not unfold alongside a narrative path in their revelation, they cannot be traced backwards to some mythical origin. Cyberrisks exist only in dissemination. Whereas virtual risks can still be traced (as objects of virtualization), cyberrisks can be mapped only because its flows are multiple.

> The map is open and connectable in all of its dimensions; it is detachable, reversible, susceptible to constant modification. It can be torn, reversed, adapted to any kind of mounting, reworked by an individual, group, or social formation . . . A map has multiple entryways, as opposed to the tracing, which always comes back 'to the same'. The map has to do with performance, whereas the tracing always involves an alleged 'competence'.
> (Deleuze and Guattari, 1988: 12–13)

The obsession in network computing with distinguishing between viruses and hoaxes can be seen as a denial of cyberrisks. Trying to trace a cyberrisk to its alleged origin in the mind of a criminal-obnoxious 'geek' may be a favoured pastime of criminal investigators and media pundits; it merely serves to mystify the nature of cyberrisk, which is never ascertained by origin but by its dissemination. Hoaxes are like viruses; they may also use email alongside television broadcasting to spread; they are as irritating and – because they proliferate autonomously from, and in fact exactly because of, the security concepts of 'safe computing' – they are as effective. The sheer amount of viruses, trojans, worms and hoaxes, combined with the speed at which they travel and the inability of virus-detection systems to stay ahead, makes 'tracing' a rather futile activity. The fact that cyberrisks can spread from and to any part of a world still carved up by national borders means that the 'policing' of cyberrisks can only work through deterrence. The interplay between the fabrication of cyberrisks and the forensics of their tracing becomes itself a strong motivator of the proliferation of risk sensibilities.

Cyberrisks thus need to be appreciated both in their singularity and multiplicity. The question 'what is cyberrisk?' is rather futile. Instead, we have to ask a different question: 'what does *a* cyberrisk do?' We could ask this question about cyberrisks only in terms of their singular perfomativity (never in general terms, nor in their particular manifestations), in particular that of their pragmatics of sense-making. Successful mapping of a cyberrisk must resist the temptation to reduce 'it' to a signifier or code. It can be sensed only in its distributions (effects). It should simultaneously be clear that this cannot be an either/or enterprise. Whenever we engage with cyberrisk reflexively, we are transcoding the matter-at-hand. Transcoding works both ways – it mediates between the virtual and the cybernetic. The reduction of 'cyberrisk' to a sign or code through signification, is thus simultaneously (in a supplementary sense) an embellishment. In the same movement, the signifying semiosis of 'virtual risks' sets into work a new information-processing *poiesis*, capable of generating an excessive induction of singularities into the symbolic order.

> Computer viruses are predisposed to escape the jurisdiction of their creators, dispersed by integrated circuitry, and capable of utilizing previously accumulated signifying material for their own explosive replication, and in these senses might be seen as the archetype of the superconductive event.
>
> (Clark, 1997a: 88)

The computer virus as the archetypical cyberrisk has been at the centre of attention of many media, not only news. Both popular science and science fiction have embraced the immense apocalyptic sensibility of cyberrisks to sketch futures totally out of control because of self-replicating artificial intelligence. Books and films about the computer age, and especially those involving computer viruses are usually framed in a dystopic setting. Like the misanthropic apocalypses of emergent viruses as nature's revenge (see Chapter 7), they tend to sketch the future in very dark colours. However, unlike the nature's revenge domesday scenario, the location of cyberrisks is unequivocally the technological culture of the west. This culture is described as dehumanizing and alienating. What characterizes many of these accounts is a connection between cyberrisk and dystopia that is mediated by intentional political violence against humanity itself. That is, cyberrisks manifest themselves in popular cultural media as the epitome of conspiracy theses.

The jouissance of viral terrorism

In *Virus* – one of the first monographs on computer viruses – Allan Lundell (1989) starts his exploration into the world of computer viruses with the case of a highly virulent internet virus that came to existence in 1988. It was designed to jam the entire system worldwide. It failed because of a programming 'error', which Lundell suggests may have been deliberately designed. What allowed this virus to be so powerful was its ability to break through the various security devices that were designed to protect the system from such nasty intrusions. It is the story of a rapidly spreading digital bug that was programmed to explore the gaps in electronic mailing systems. As a peculiar twist to this story, Lundell adds a genealogical note: the designer of this virus, Robert Morris Jr, is the son of a world famous expert in computer security, Robert Morris Sr.

To program such a virus is both seen as a criminal act and a highly skilful accomplishment – the aesthetics of the sublime are those of a radical evil. The passion that both father and son Morris have for computing is a libidinal economy that is simultaneously inscribed in the paranoid warfare economy of the state apparatus (Doel and Clarke, 1997). Lundell finds a closure in this story by pointing out that Morris Jr's act has led to a wealth of insights in the vulnerability of networked computing systems as well as in the operations of such skilfully designed viruses (Lundell, 1989: 13). The anxiety of risk of failure is thus offset by a will to know and a belief that complete security and regulation can be achieved.

However, cyberrisks are not that easily put off by Oedipal desire as they operate in a rather different libidinal economy. Since Lundell's now very dated book, our world has become increasingly dependent on computing systems and telematics. In the mean time, a plethora of new viruses have appeared, most notably 'Michelangelo', 'Melissa', 'I Love You', 'Red Alert' and 'Nimda'. The most feared viruses are those that use email attachments, appropriating a device designed to make telematics more user-friendly. With the global domination of Microsoft over the design and standardization of software applications, telematic systems have become extremely vulnerable to cyberrisks because of the relentless drive to acceleration and increased compatibility. Indeed, nomadic forces of cyberterrorism find it easy to travel through the deterritorialized spaces of global capitalism.[7]

However, for every virus, there have been hundreds of hoaxes. Perhaps one of the most famous hoaxes has been the 'Millennium Bug' or Y2K. During the mid-1990s, noises emerged, primarily in the form of rumours, that computing systems could experience a breakdown at the turn of the millennium because of the way in which computers kept dates. The year was marked by only two digits, to save memory space on the already overcrowded original memory boards, a practice that was simply sustained even when no longer necessary. At the year 2000, when those digits turned to 00, all hell would break loose, since computing systems would act as if it was 1900. Billions were spent on rewriting the configurations and replacing computing systems with ones that were Y2K-compatible (Schwaller, 1997).

Then, when the clocks turned to 00.00 on 1 January 2000, nothing happened. Even computing systems that had not been upgraded, such as those controlling Russia's nuclear plants, functioned 'normally'. Instrumental fidelity did not fail us, and the world breathed a sigh of relief. The point however is not that it was a hoax, but that no one, not even computer experts, was able to either verify or deny the rumours. It also showed how badly prepared our institutions actually were, and how slow they reacted to such potentially lethal risks. Above all, it showed how deeply entrenched microchips and data-processing have become in our daily lives. People were invited to make connections not only between the hardware of their PCs, cars, video recorders, televisions and refrigerators, but also between their own private systems and those of a collective nature such as banks, traffic control, hospitals, public transport, aviation, nuclear plants, policing, etc. And this had to be thought of as on a global scale. Hence the situation at the Russian nuclear plants was as significant as whether we could still draw money from a cash machine. Indeed, people were encouraged to reflect on the global complexity of connectivity and to see how fragile they have become in the face of risks. This is no hoax but a truly transformative event in contemporary technological culture.

Now that the smoke has cleared, everything seems to retract rapidly into the former segregated conceptions of modern life. Apparently, no lesson has been learned. Technological culture has indeed been shown to be quite resilient. The adhocracy of the quick-fix still rules supreme. Yet, a sense of global interconnectedness has not entirely disappeared. Internet viruses occasionally strike at the heart of telematics, and point towards its inherent precariousness. Internet security has continued to be a top priority for governance, commerce and the military, and initiatives are being taken to set up a worldwide information regulation policing system; something currently carried out de facto by the Federal Bureau of Investigation (FBI) and – since 11 September 2001 – by the US Army.

Risk remains associated with anxiety and paranoia. In the world of telematics, cyberrisks are particularly uncanny in the way in which they seem to bring to life medieval sensibilities about the dark side of human being. Such ideas can be found in two relatively recent but not quite so successful films: *Cyberjack* starring Michael Dudihoff; and *Cybertech P.D.* starring Lorenzo Lamas.[8] In *Cyberjack*, a multinational computer company is involved in the development of a computer virus aimed to defend computing systems against all kinds of unwanted intrusions. The company is then taken over by a group of terrorists who seize the virus to create chaos in order to control the world. Michael Dudihoff, the hero, manages to kill them all and save the world.

The film opens with the following quote that is allegedly Stephen Hawking's:

> A computer virus should be considered a form of life; but I think it says something about human nature, that the only form we have created so far is purely destructive. We've created life in our own image.

Although, in the narrative, risk is predominantly framed in terms of 'bad' people seizing a telematic system through a malicious computer virus, cyberrisk constitutes the very possibility of this taking place. As a form of life, the computer virus had already enrolled so-called cyberterrorists to engender its autopoietic take-off. Similar to the conducive role played by viruses in genetic engineering, computer viruses are a central force in telematic systems, they are the conductors of transcoding, moving from one system to the next, rewriting the codes as they move along. It is thus no wonder that cyberrisks are a source of great concern, even panic. They signify the limits of our technological culture through the excessive nature of their pathogenicity. However, because they are often associated with and even reduced to specific wilful individuals, who are motivated by evil, cyberrisks are also recuperated by a morally coded tech-noscience of information forensics and the governmentality of information policing. Like Dudihoff's bold hero, these agents of modern technological cul-ture may rescue us from our damnation. Because cyberrisks are often denied any agency apart from their creator's – that is because we fail to understand cyber-risks as cyberrisks – we promote a false sense of security and endorse a misplaced faith in securing and regulating systems. Nowhere has this become more clear in the events running up to the attack on the World Trade Center and its after-math, for example in the anthrax scare of October 2001.

Whereas in *Cyberjack* the threat comes from 'bad guys' using 'neutral' tech-nology, in *Cybertech P.D.* the threat is technology itself. Virtual reality (VR) is placed on the same continuum as pornography and drug use. The key word here is 'addiction' – the institutionalization of mental and physical dependency upon external stimuli. *Cybertech P.D.* is thus based on a more traditional police narrative and involves the usual mixture of sex, drugs and violence. It features a cyber-policeman played by Lorenzo Lamas who is on a quest to arrest a former drug-cartel baron, now VR-entrepreneur. The emphasis here is not as much on viral cyberrisks but on a general sense of corruption and violation that virtual realities engender, in particular when bio- and cybertechnologies are intercon-nected.

In *Cybertech P.D.* the hero-cop is technologically enhanced and con-nected – via a Gibsonesque sort of sim-stim (simulation stimulant) – con-nected to the Cybertech police force. However, his heroic status solely depends on his ability to operate outside the technosphere. He must remain an individual and resist the seductions of virtual sex and cyberdrugs. His moral code stems from his romanticized integral-corporeal embodiment (humanism). Hence the real threat comes from the cyber-bourgeoisie-mafia who have now seized upon biotechnology to clone actual human beings (mostly women of course) and turn them into sex objects. If this transgression of the virtual is accomplished, there are no safe havens for human corporeal integrity – all is corrupted. In the climax, the hero injects himself with a superdrug to overcome the impossible, defeats the bad guys but seems to be lost forever, as he himself is now corrupted beyond repair. However, through embodied desire for natural romantic love, he is 'rescued' by his female counterpart and able to resist technological corruption, thereby re-establishing Human Nature over (demo-nic) Techne.

Addictions are cyberrisks because they work as a virus. They engender a parasitical dependency relationship in which agency is surrendered to an exter-nalized force. The integral body is corrupted and foreign agents take over. In medieval times, such phenomena were often associated with demons; angels of the Evil One, who would enter a body, often through sin, and take control. The term demons is derived from the Greek *daimones* (Pagels, 1995: 120). St Justin, an early Christian thinker, took this Greek term, which at that time still had a neutral meaning referring to spiritual energy (similar to, say, the Maori notion of *Hau*, referred to by Mauss (1990) in *The Gift*) and associated it exclusively with the powers of Satan. It is not surprising that addictive behaviour was often seen as demonic. Addicts act as if possessed by an external force, to a degree it becomes self-destructive. The loss of personal integrity is also coupled with the loss of free will and the ability to make correct judgements.

Addictions prominently features among cyberrisks of the 1990s. Varying from addictions to games, chat rooms or internet pornography, ICTs are often demonized in the sphere of leisure and consumption. Hence, a recent outcry over paedophiles using internet chat rooms to contact underage teenagers is merely an example of the way in which virtual reality is conceived as an inherently dangerous and easily corrupted space, situated beyond morality and security. This anxiety goes beyond the fear of paedophiles obtaining virtual presence in the safe haven of a child's bedroom; it extends into the fear of the child her- or himself being corrupted by it.[9]

Both *Cyberjack* and *Cybertech P.D.* display a deeply modernist neurosis as their founding psycho-semiotic constituency. Both express an anxiety over the vulnerability of the human condition in the world of cybernetics. They thereby echo a whole genre of dystopian films about cyborgs and cybernetics, such as *Terminator*, *Robocop* (and their sequels), *Strange Days*, *Bladerunner*, etc., which all evolve around an association between telematics and pathologized human evil, in particular megalomania (Holland, 1995). Holland is right to point out that a fascination with cyborg films is often induced by paranoia, but Nigel Clark's (1997a: 79) assertion (derived from Baudrillard) that we are actually fascinated by the creation of a catastrophe should equally remind us that not all apocalypses are dystopic, and not all dystopias are anticipated by paranoia.

Obviously, such films, like Lundell's book, are not outside the genre and ethos of the socio-cultural climate in which they were engendered.[10] They should thus not be seen as anything but representations of our own relationship to the present, mediated via a construction of 'futurity'. That viral terrorism and biotechnological corruption threaten the integrity of the system-as-we-know-it is left unproblematic; it points to a fundamental psychopathological disorder of modernity itself: neurosis.

Lundell's *virus* is stifled with this modernist paranoid neurosis. However, his accounts of the various viruses, worms and trojans that plagued the computer world in the 1980s also include a fascination with destructive evil and the ability of peculiar cybercowboys to trace and neutralize them. Moreover, his book offers itself as a *pharmakon*, holding both the poison and the medicine (Derrida, 1981). Indeed, like a television evangelist, Lundell paints both the apocalypse and the possibility of redemption ('buy this book'). The fascination with the 'preprogrammed catastrophe' constitutes the enigma of his encounter and becomes further apparent through the numerous hints at the genius of particular virusprogrammers. For Lundell, viral design involves a kind of narcissistic aesthetics – computer nerds create viruses to show how clever they are by testing their masculine egos in a new and powerful environment. The quality of the virus is interpreted in terms of the effects it has on users, in particular its velocity and degree of infectiousness, or 'epidemic of simulation' (Clark, 1997a: 78).

Thus, Lundell leaves us with some rays of hope – there are remedies against computer viruses. During the late 1980s and early 1990s, many network operators tried to cultivate an ethos of 'safe computing' – often in close analogy to the AIDS-hype which by then had reached its peak. Not buying illegally copied software or copying software from message boards,[11] never using disks or opening email attachments without checking them for viruses, installing virus scanners on mainframes and PCs, all were part of a newly institutionalized anti-virus campaign. Yet, despite all these efforts, many systems proved highly vulnerable to viral epidemics. For example, the computer network of the University of Wales in Cardiff experienced three main outbreaks of the 'Trajectory' virus between March 1995 and March 1996. By then, viral scanners were already widely used and installed to automatically monitor all the servers on the network. Apparently these measures were insufficient as the virus was too quick and too virulent, too subtle and too effective, for such institutionalized safety-nets to function properly.

Of course one may attribute such pairings of apocalypse and redemption to a genre of modern fiction-writing; but, as has become clear since 1990, it is the scientists who are now the main contributors to the panic scenes of risk society. Nuclear technology, genetics, waste management, AIDS, antibiotics and computer viruses are all incorporated into stories written in anxiety and fear on the one hand, and an optimistic faith in an ultimate victory of reason on the other hand. As the apocalypsis starts to gain momentum, reason becomes increasingly revolutionary, that is, the 'solutions' become increasingly 'unreasonable' (Doel and Clarke, 1997; Van Loon, 1997a).

To reduce these apocalypse/redemption pairings exclusively to forms of virtual risks, that is to forms of popular cultural expressions that have no index of effectivity outside 'the' mediascape, is to miss a vital point: they have an innate sense of en-presenting themselves beyond the technologies of visualization and signification upon which virtual risks rely. It is the connectivity between the desiring machines of the apocalypse and those of the unreasonable that engender the cyberrisks we are subjected to today.

Neutralize the neurosis!

What is so typical of the risk society as described by Beck (1992) is that it suffers from a deeply entrenched and very modernist form of neurosis. The fate of modernization is that of unforeseen and unforeseeable consequences that turn against the movement itself. Risk society marks that moment when such a fate is recognized as imminent to the systems that are designed to control and overcome it (mistaking fate for destiny). Why, then, do so many (social) scientists discuss the ecological fate of risk society in terms of institutionalizing risk management? Why, then, is so much energy invested into developing a new rationality to counter the problems of modernization, for example in terms of 'sustainable development'? Sustainability may be something that has less to do with management and control, but instead with letting go, with breaking down the artefacts and barriers within which the state apparatus has entrenched itself against attacks from the margins. Perhaps the most vital lesson can be learned from the viruses themselves. They tell the story of the eternal return.

In a world in which the survival of the human species is no longer a certainty – due to a seriously disrupted ecosystem as a result of which they are no longer capable of providing enough oxygen, water, carbohydrates, protein, vitamins and minerals without lethal intoxications or infections with encephalopathic prions and such – we may not want to wait for institutionalized reflexivity to pull us out of the sewer in which we find ourselves stuck while busy creating it. The media–science connection has made this all too clear – we are doomed, at least for the time being. Politics, law, governance and commerce are not going to rescue us, as their timescales are far too short to deal with the organization of the risks and hazards they effectuate (Adam, 1998, 1999; Van Loon and Sabelis, 1997).

Because the apocalypse looms large, we have to turn to the revolutionary desire of the unreasonable, the aesthetics of the nomadic war machine. This is the lesson of the cyborg that Donna Haraway (1990) points us towards in her manifesto. The command, control, communication, intelligence (C_3I) systems of the state apparatus can be turned into a nomadic war machine if they are reconfigured as an assemblage of monstrosity. The nomads living beyond the borders of 'the system' can connect and constitute formations in border wars against the attempts to solidify everything in institutional forms and networks. Hakim Bey (1991) pointed towards such formations in terms of temporary autonomous zones; their temporal autonomy is the greatest strength and the greatest weakness of the nomadic warrior tribes. Connections with the state apparatus are necessary to interrupt and disrupt the flows of information, money, goods and people; the force and energy of the C_3I systems are needed to have any effect; however, as it connects to the apparatus, the nomadic war machine becomes part of the apparatus, infected by its desiring machines and self-valorizing neurosis, it becomes a corrupted change agent, fixing solutions until it is part of the problem.

If we now read back the opening citation from Lyotard, we can perhaps see how cyberrisks allow us to theorize mediation in a far more radical way than as merely a representational practice hovering between the politics of containment and of moral panics. It extends from Heidegger's reflections on the essence of technology as a mode of revelation (Chapter 5). In a similar vein, Lyotard's assertion allows us to distinguish this essence of technology from technology itself. Technology used in this way merely refers to the information-processing capacity of material systems applied to sustain themselves. Read in this order, it becomes clear that virtual risks are thus particular manifestations of cyberrisks that allow us to make sense, to trace an origin, to place them in a field of signification, to subdue them to our discourses, to invest them into the organization of our precious symbolic order. The insurgent telematic symbiosis is merely facilitating their transcoding into the social.

Hence, if we stop thinking of media as representational apparatuses and instead adopt the model of media as signifying cybersystems, we may be able to understand the consequences of mediation as 'materialization'. When the sign crosses over from semiotics into pragmatics, it becomes electrified, leaving the visual, entering the tactile, setting into work a range of connected machines. This is not to say that there is no point in studying media representations of risks and the environment; however, these must be developed in a more immediate materialist-existentialist sense. For example, the point to make about the media involvement in the BSE crisis in the UK is not simply or even predominantly that of misrepresentation and the foregrounding of nationalism (in terms of economics and politics) over issues of public health and moral integrity. Of course, these observations are important; but to exclusively focus on them is to overlook the more fundamental issue of how media, science, governance and commerce came to be interconnected in the first place. Only by focusing on the exclusion of 'an outsider-logic', that is a logic that is not part of 'royal science' and its assumed certainties, could the question of 'integrity' be shifted swiftly from that of the living individual, but dementing bio-cybernetic body, to that of the abstracted social body. There is no conspiracy at work here; it is simply a matter of systemic (while hesitating to say 'institutional') closure – the currency of the symbolic exchange was that of modern neurosis, not that of embodied spirituality or exuberance. What was turned into a sponge was not the human brain, but the system itself. This, truly, is panic ecology (Clark, 1997a).

Again, we need not follow the ideological apologetics of modern warfare here, and should extend the notion of war to all forms of collective violence and civil unrest.

In this chapter, we seek to apply the argument of risk as a turning in technological culture with reference to an emergent apocalypse in the form of 'urban warfare' in the heart of modern civilization. The chapter thus provides the fourth illustration of the theoretical framework outlined in Part I. It focuses on the particular role of media in the organization of violent 'crime' and in particular urban disorders. It will look at the role of media technologies, technologies of policing and judicial technologies in the generation of 'accounts of violence'.

In previous chapters, we have seen how media are actively engaged in the politics of waste management, the engendering of pathogen virulence, and central to the inauguration of cyberrisks. Through fiction, film and television, media enable a wider dissemination of significations of risks, and an enrolment of more distant actors into the risk-management networks. However, especially in the previous chapter, we have seen that the role of media is not simply that of giving functional support. In cyberrisks, the media themselves are becoming pathological. In this chapter we shall continue this theme of the ambivalence of risk mediation. Focusing on, first, the case of the beating of Rodney King, second, the trial and the verdict of the four White police officers who were videotaped beating him and, third, the broadcasting of the beating and verdict and the subsequent 'LA riots', it will be argued that the risk of collective violent disorder is endemic to the media technologies with which the beating, the verdict and the broadcasting of both were actualized. This will be used to open up a more general discussion of the changing nature of 'ordering' in a turning of technological culture to the warfare state.

The concept of warfare state thus applied suggests that the breakdown of social order is never far away in a risk society, not only because the catastrophic potential of technoscience is no longer containable by the insurance principle (Beck, 2000), but also because the very technologies that allow us to make sense of risks, and that cultivate our risk sensibilities, have been affected by the virulence of these same risks. The constitution of particular actor networks between (social) science, governance, media, commerce, law and military reveals that the combined visualization, signification and valorization of violence produced by this assemblage of the warfare state, while trying to limit the monopoly of legitimate violence to the state, cannot but proliferate its virulence because of its fundamental arbitrariness. This is what so many ethnicized and racialized 'minorities' find out *vis-à-vis* the state's 'interior' policing force (including immigration, customs and even social services). This is also why these marginal groups are usually (but not always) the first to be engaged in 'guerrilla warfare' against the state.[1] This warfare can take many forms, from organized crime to international terrorism to individual acts of 'terrorism' such as that of the Unabomber (Parfray, 1990). They usually engender other actor networks, developing their own assemblages of technoscience, governance, mediation and commerce, which – by their very nature – also remain part of dominant constellations.

The role of media is central in these assemblages. As Thompson (1995) and Dahlgren (1995), following Habermas, have argued, the public sphere of our modern society is to a large degree arranged and ordered by mass media. This also has profound implications for the way in which the state organizes social order. It has replaced much of the politics based on physical space with those based on 'virtual' space. Police presence on the street, for example, is now supplemented with CCTV systems, creating a panopticism outside the physical enclosures of total institutions. Through ever more sophisticated (sub)molecular forensics, in particular DNA-sampling, policing now inhabits a physicality beyond that of the 'common' empirical senses of human being. This is how the society of discipline transforms into a society of control (Deleuze, 1992).

The supplement of spatially framed body politics and street politics with those mediated by media technologies (displacing the physical to the virtual and molecular) is deeply affected by the politics of risk aversion. For example, by stressing a range of crime risks, police forces in western society have been highly effective in increasing access to technoscientific innovations in terms of resources and legitimacy. The politics of risk aversion, however, only work by transforming physical violence into symbolic violence. That is to say, by displacing the main frontline of the state of warfare from the street to the laboratory, the monitor and the data processor. This is never simply a sublimation of bads into goods. The 'containment' of risks returns to the streets, and to the embodied physics of violating and violated bodies. This chapter shows that risks engender assemblages in which humans, technologies and spirits are endowed with agency. However, its operative logic can never be reduced to the sum of motivations of actors. Instead, there remains a beyond, something larger, a monad, that reveals itself in its absence, as the will-to-power, the will-to-organize, the will-to-control.

Moral panics

All risks, even those specifically (but never exclusively) related to the environment such as waste or emergent pathogen virulence, have a social and cultural embedding and are inextricably linked to the workings of technologies of mediation, which, through making visible, attributing meaning and allocating value, enframe and reveal an extended gap between possible and actual calamities. This gap is the essence of the virtual objectivity of risks.

Whereas the virtual nature of risks is immediately clear in the new digital media of telematics, in this chapter we will explore how the concept of cyberrisk may also be applied to reflect on 'old' analogue media such as television and video cameras. Using McLuhan's (1964) famous reflections on television as an externalization of the human central nervous system, we discuss the way in which media technologies have enframed and revealed the phenomenon of collective violence, or 'urban riots', as a spatially diffused force. The meeting

172 The Four Riders of the Apocalypse

of media may sometimes function as a rupture in the relatively smooth flows of social ordering, and engender a 'turning' in which the truth reveals itself, if only in a flash of lightning, before being recuperated narratively and mythically in the realigned discursive formation of common sense, which consequently becomes more unstable.

The risks of such media-technological revelations are of course not limited to just collective violence but have possible implications for all kinds of phenomena that have been the subject of 'moral panics'. However, we shall see that they have particular potency when enframing/revealing as pathogenic certain strangers in our midst, such as 'youth' and groups marked by ethnic and racial difference. The collusion between proximity and difference in the formation of 'pathologies' is what propels particular social risks beyond human and institutional control. This enables such risks to replicate themselves almost infinitely and instantaneously, merely reappropriating the media technologies that set them into work in the first place.

More specifically, this chapter will focus on the role of media technologies and risk in the pathologizing of 'race' through the formation of moral panics around 'urban riots'.[2] The argument pursued here is that *mediation* is the primary practice/process of violent identifications which constitutes rather than uncovers the logic of racialized violence. It is through mediation that racial matrices are imposed on the risk-inducing *sensibility* of such events as urban uprisings. The analysis evolves around the case of a videotaped beating of a Black motorist Rodney King by four White Los Angeles police officers and the subsequent acquittal of these officers in court by a nearly all-White jury.

The argument is based on a semiotic analysis of the formations of 'race' in the television news coverage of the 1992 Los Angeles 'riots' that immediately followed the acquittal (Van Loon, 1996c). Focusing on denotation/connotation, intertextuality, chronotopicity (also see Van Loon, 1997b), narrativity and discursivity, the objective of this study was to trace the racialized identity formations that emerged as the giving of an account of 'what happened' in the form of TV images, sounds and texts. What appears to be the central focus of most news coverage is the function of 'race' as a device that has been mobilized to account for the phenomena associated with rioting (for example as Black-on-White violence) and to organize explanations of the causes of such rioting.

The relationship between the acquittal and the Los Angeles riots has generally been perceived within news broadcasts as a shift from the world of opinions and ideas to that of brutal physical force (Butler, 1993; Fiske, 1994; Gooding-Williams, 1993b; Swenson, 1995; Van Loon, 1997b). The symbolic order in which this shift has been embedded, however, is deeply racialized. The racial element involved in the transformation of outrage (symbolic violence) to (physical) 'violence' becomes more clear perhaps when taking into account the reversed process: the transformation of the physical violence of the beating of Rodney King into the symbolic violence of the acquittal of those responsible for the assault. This justification of violence is part of the same aforementioned race struggle, because it was performed by a nearly all-*White* jury to clear four *White* policemen. Both police and jury are 'authorities' that function in an imagined community in which 'Whiteness' is the norm. In the force exercised by these authorities resides a reconstitution of this Whiteness by asserting that such violence is legitimate when it is used to *reinforce* law and order, i.e. the norms. Risk has played a central but generally neglected role in this process.

Racialization through mythification

Critical analyses of media coverage of riots have revealed that the televisual mediation of riots often involves a *mythification* of rioters which evacuates the *sense* from rioting and replaces it with meaning formations that are discursively pre-constituted and therefore taken-for-granted (Fiske, 1994; Gooding-Williams, 1993b; Keith, 1991, 1993; Lewis, 1982; Solomos, 1988; Swenson, 1995; Trew, 1979; Van Dijk, 1991).[3] Mythification turns race into an essence and negates the very construction this entails. One particular element of mythification that is often discussed in this context is the symbolic association between violence and race; that is, violence is identified within a racial matrix. As Houston Baker (1993: 40) has argued, violence is always-already racialized, for example in the silence of the slave before 'his' master. In the television news coverage of the 1992 Los Angeles uprisings it returned in two forms: as a lack of law and order and as a lack of (economic, political, social, cultural, racial) integration.

Whereas Baker and others (e.g. Butler, 1993) do recognize the central role played by 'lack', this is primarily understood within a psychoanalytic framework (e.g. Fanon, 1952/1986). Consequently, there is a tendency to essentialize the association between these types of 'social' lack and lack in terms of 'racial impurity' and to reduce them to the Oedipus Complex of the white man. Apart from turning 'race' thereby into a substitute category for masculinist sexual frustration, it also overly reduces the historical complexity of racialization to a mere psychological pathology. Instead, it may be more fruitful to historicize this notion of 'lack' by invoking the concept of risk. That is to say, racialization is a historically specific response to a modern western conception of risk associated with particular formations of belonging and violence that are mythologized as hierarchies of purity (Goldberg, 1993).

The association between race and violence as a *lack of law and order* appears for example in accounts of rioting as expressions of a lack of concern for property and people, opportunist gang-related crimes and racial antagonisms. These associations of race and violence belong to what Justin Lewis (1982) has termed the *law and order discourse*. In this view, violence is the equivalent of disorder, a lack of order, whose causal structure can be attributed to race itself or, more precisely, to racial difference – the difference of race. As Young (1994) has argued, race invokes a discourse of variable degrees of impurity. Race is thus understood as a lack of purity. The lack of purity manifests itself in a disregard for the institutions of modern western civilization, especially notions of property and proper conduct. This, however, is a typically modernist conception of impurity; one that testifies not to a universal human desire for racial superiority, but to the very invention of a language of distinction to naturalize the historical particularity of one's universalistic claims (Gilroy, 1993), in this case of 'man's' domination over nature, the divine and history itself.

The risk of 'race', then, is a risk similar to that of waste or emergent viruses; it is the risk of abjection. As Douglas (1966/1984) has argued, a social system under increased external pressure will tighten its control over boundaries. By taking 'his' destiny in 'his' own hand 'modern man' has imposed onto 'himself' a strain that never eases. Being both origin and destiny, 'modern man' cannot but become a neurotic paranoid as everyone envies 'his' achievements. Racialization is a strategy of boundary maintenance, and risk signifies the dangers of boundary violations.

The pertinence of race in accounts of violence as a lack of order can even be traced in explanations of riots which have no clearly articulated racial structure. For example in the disturbances on the Marsh Farm estate in Luton (7–9 July 1995) and in the Hyde Park area of Leeds (11–12 July 1995), where 'youths' were mainly held responsible, the causal attributes were predominantly sought by news reports in the 'incompleteness' of these youths which made them susceptible to 'criminal tendencies'. This incompleteness was not accounted for itself, but treated as a cause that presented itself in a lack of respect for law and order. Although not racialized in terms of skin colour, the racial matrix of modernity was clearly present in the symbolic association between incompleteness, risk of violence (boundary violation) and a lack of respect for law and order. This association is prolific in descriptions of 'rioters' and 'criminals' as a specific *breed* (type, stock, *race*) of people. This clearly represents a discursive strategy of naturalization through the invocation of myths.

Closely related to the aforementioned law and order discourse is the second prominent discursive formation to which journalists and pundits often resort to account for racialized violence in the television news coverage of disorders. This formation pictures disorders as conditioned by socio-economic deprivation; social, cultural and political marginality, the breakdown of families, communities, and societies, racism, and a lack of racial integration. These can roughly be grouped together in what Michael Keith (1993) has termed *social problems discourse*. Here violence is the equivalent of *lack itself*, that is a lack of social, cultural, political, economic, racial *integration*. Explanations and interpretations that draw on this type of discourse always deliver their accounts as attempts to uncover deeper, often hidden, causes of rioting. In contrast to law and order discourse, advocates of social problems discourse thus reject the assumption that rioting phenomena can be understood as such and instead offer a symptomatic reading of such events as the effects (appearances) of other forces. Instead of a lack of respect for law and order, lack is addressed in terms of socialization, integration or equality. However, racialization is as prominent as in the first type, since within the historicity of modernization, there is an emphasis on an association between racial difference and differential civilization and cultural values (Goldberg, 1993; Miles, 1989). Racial difference is thus the equivalent of a *lack* of culture, civility and socialization.

Again the impurity of race manifests itself as 'lack', this time not through a naturalization of racial difference, but through a naturalization of the social system itself. Modern society is seen as evolving through integration, incorporating elements of difference. These elements of difference may initially 'lack' the necessary qualities to become fully functional but have to potential to become the same. Whilst, in this view, racialization is merely a temporary phenomenon that will disappear over time, it strongly relies on a notion of 'social risks' as corresponding with racialized difference. In order to effectively contain those risks, racialization becomes a necessary evil of population control; the identification of special needs enables agents of the state to legitimize greater capacity to intervene in the everyday lives of those identified as such. Risk justifies such interventions as well as the racializations through which they are being applied. Although the allocation of 'blame' in relation to this lack/risk may point to the state rather than the racialized other, and this may lead left-wing liberals to justify their critical involvement, the targets of the applications that follow such significations are invariably these groups on the margin.

Once made the equivalent of lack, (risk of) violence is mobilized as a discursive object to *make sense* of these disorders as a negativity that must be countered, both in terms of the legitimate force of the authority (state) and the symbolic force of 'meaning-fullness' upon which the media principally operate. When law and order discourses proliferate, the 'combat' of lack (control of risk) is likely to be more coercive and repressive than when social problems discourses are more prominent (Keith, 1991, 1993; Lewis, 1982; Solomos, 1988). Although this difference is certainly significant in political terms, it takes place on an identical fundamental assumption: racial difference equals lack equals (risk of) violence. This *logic of equivalence* engenders an identification of violence as a designated set of practices associated with race in very particular terms. However, because intricately linked to 'lack', the relationship between violence and race remains itself rather elusive. It is invariably subject to mythification and thus naturalization. The implicit invocation of risk enables the governance of social order to displace race. Whereas the concept of 'race' may have originated in modernity, racism does not sit comfortably alongside the ongoing universalizing pretensions of modern thought (e.g. progress). It is a reminder of the fundamental inconsistency of humanist reason.

Moreover, the logic of equivalence operates on the assumption that it is possible to separate the violence that is associated with disorder from the violence that is mobilized to restore order. To be more precise, the logic of equivalence – which is exercised through risk – grants itself the capacity and legitimacy to distinguish illegitimate from legitimate violence. However, it is the chain of equivalences between 'race', violence, disorder, crime, pathology, excess and risk that attunes the warfare state to mobilize its resources towards the control and even elimination of difference. Through the concept of 'risk', the warfare state is able to recode the territoriality of public and private spheres, to colonize the domains of law and justice, to link with governance, technoscience, media and commerce in the mobilization of force. The coupling of violence and risk enables the warfare state to increase its strength and authority through the anticipation of breakdown of ordering.

The beating of Rodney King

A good way to problematize the relationship between violence and racialization in the context of media discourse is to look at a series of events following the beating of a Black motorist, named Rodney King, by four White Los Angeles police officers on the evening of 3 March 1991. Rodney King was stopped by officers of Los Angeles Police Department (LAPD) highway patrol after a high-speed chase. As soon as he was forced out of the car, four LA police officers took over the responsibility of the arrest by delivering 56 blows to Mr King's head, body and limbs in a beating that took 81 seconds before he was handcuffed and taken to hospital with severe injuries to his head, body and legs. However, the officers did not know that this beating was accidentally videotaped from a balcony by an amateur cameraman who was trying out his new camera (Baker, 1993; Fiske, 1994; Swenson, 1995).[4] As soon as he discovered what he had recorded, he went to a local police station where he was told that the police were not interested. He then went to a local TV station which immediately broadcast the tape in full. As we now know, the rest is history.

When the prosecution brought charges against the officers, it was thought that the trial should not be held in Los Angeles itself because of possible bias against the defendants, and the venue was subsequently changed to Simi Valley, a 96 per cent White suburban city to the north of Los Angeles in Ventura County. The prosecution presented the issue of the beating to an almost White jury (ten out of twelve; the other two were a Korean and a Latino) under the format of the question whether the beating was a case of excessive violence (as the prosecution argued) or reasonable force (as the defence argued). The result of the trial is well known. The jury decided the videotape showed that the officers were using reasonable force under the collar/colour of authority and acquitted the policemen on all but one charge.

The extraordinary aspect of the case was not the beating, but the fact that it was videotaped and that the videotape was broadcast, nationwide and indeed worldwide. The violence that was being displayed seemed real beyond any reasonable doubt. The jury in court, however, judged differently: it was not an act of violence but a display of authorized procedure. The difference is conceived as a question of legitimacy: violence that is legitimate is not identified as violence but as reasonable force. The task facing the media after the trial was to attune the intuitive truth of the videotaped images (the policemen are guilty) with those of the prevailing judgements (the policemen are innocent). Hence, the question that most pundits raised immediately after the trial was 'how' this verdict could be possible? Did we not see with our very own eyes that the beating was excessive? What we thought of as *seeing*, i.e. excessive violence, was interpreted as 'reasonable force'. How can we explain this rupture between 'our' intuition and this verdict?

The verdict

Although much has been written about the way in which the jury achieved a reading of the videotape that led to the acquittal of the four police officers (Baker, 1993; Butler, 1993; Gilmore, 1993; Fiske, 1994; Gooding-Williams, 1993a; Lipsitz, 1994; Williams, 1993), it is necessary to repeat some of the analyses to understand the complex mechanism by which the jury came to interpret the videotaped beating as an enactment of reasonable force, and the role that can be attributed to the racial matrix of 'identification'.

First, it must be made clear that the jury never appropriated an explicitly racist stance. They never justified the beating simply because Rodney King was Black. Such a form of overt racism is clearly unacceptable in modern western society. However, whereas 'race' was removed from the explicit interpretation of the videotape in court, it returned implicitly as the repressed of this very interpretation. What effectively took place was a whitewash not only of the racism behind police brutality, but also of that behind the verdict itself, and – subsequently – of the US justice system as a whole.

The acquittal of the four police officers was the result of a reinterpretation of the videotape, following three steps. First, the relevance of the videotape was simply denied. Instead, the argument of the defence, taken over by the jury, was that 'these police officers have a job to do' (ABC *Evening News*, 30 April 1992). Second, it was argued that the videotape did not tell the whole story. It missed out events that took place before the camera started rolling; the images – which were in black and white and blurred – were simply not clear enough to show what actually happened. It was noted that 'several jurors said they were persuaded by the defence argument that the videotape did not tell the whole story and that Rodney King represented a threat to the officers' (ABC *Evening News*, 30 April 1992). Butler (1993) presented an interesting analysis of this reversal of risk. Rather than Rodney King being at risk from police brutality, the police officers were at risk from this dangerous (and allegedly criminal) Black man. This reversal is possible only if the Blackness itself already represents risk, which – as we have seen – it exactly does by virtue of its incompleteness.

The third step entailed a return to the videotape. Although the tape may have been an incomplete piece of evidence, it was completed by the testimony of the officers. Butler (1993) persuasively demonstrates how the defence (and the jury following their lead) interpreted the images of the beating as the foundation of the acquittal. She argues that the video was recontextualized in a White-visual paradigm to show that Rodney King was resisting his arrest, that he was in control of the entire situation until he was finally forced into submission, and that he therefore represented a risk to the officers involved. As Fiske (1994: 132–3) has argued, this *logorationality* was enabled by the various technologies of visual transformation and hybridization of the video images.

Fiske defines logorationality as the production of particular forms of knowledge on the basis of the power of 'the' word.

Evidence from the trial itself shows that the jury was more than susceptible to a carefully selected set of interpretations of the videotape, in particular those put forward by the defence attorneys and sergeant Stacey Koon. In other words, the videotape was not denied input, but was operationalized to provide exactly the 'evidence' of the officers' 'innocence'. For 'the likes of Rodney King', the tape thus became itself a technology of repressive racialization, mobilized to provide a foundation of 'reasonable force' induced by risk. In asserting that the officers used 'reasonable force', the jury thus actively sought to rationalize force exercised in the name of (White) Americanicity to conceal its primordiality under a veil of universalistic legitimacy. Hence, the 'truth' established through a logorationality of legal interpretation exposes the very *madness of reason* of a legal-technological culture (expertise) and the way in which it draws upon the concept of risk to justify itself.

Media hybridities

The madness of reason underpinning the acquittal clearly shows that the video-tape, which was brought to trial as the main witness for the prosecution, turned out to be a major witness for the defence. Many critics have pointed out that *seeing* itself cannot constitute an account (Butler, 1993; Crenshaw and Peller, 1993; Fiske, 1994; Gilmore, 1993; Gooding-Williams, 1993a; Lipsitz, 1994; Swenson, 1995). Indeed, since vision is already embedded in particular discursive locations, the empiricist notion of reality as self-evident must be refuted. Images do not speak for themselves. However, this should not lead us to dogmatically deny all 'reality' to the videotaped beating. Albeit based on an all-too-easy acceptance of the illusion of transparency with which electronic media endow an event (Vattimo, 1992), intuitive reflections are not therefore without value. The critique of realism does not square with the strong intuitive *sense* that the videotape *does* show excessive violence under the collar/colour of authority. The imperative not to deny intuition any real force is perhaps best expressed by an African American woman featuring in ABC's main evening broadcast on 30 April 1992 saying: 'They told me what I saw on television did not happen and I am infuriated that they would think I am ignorant enough not to know what I saw.'

The aim of this section is to induce from this anger an agonistic analysis, or 'angry writing' as Michael Keith (1992) once called it. The problematic rupture between intuition and judgement regarding the beating of Rodney King and the acquittal of the four police officers must be located at the level of mediation itself. It is based on the prosecution's failure to understand the medium of video recording and the way it had been appropriated in court (for a similar argument, see Fiske, 1994). At the root of that failure lies the very notion of *judgement*.

Any claim to identify violence is based on the assumption that judgement always comes before violence. It is the failure to conceive judgement as itself symbolic violence. In the case of the verdict, the jury, who were assigned to the special function of judging the event, exercised symbolic violence in the form of what Lyotard has called 'the terror of the sublime': an aesthetic of rationality. The violence of judgement, however, also emerged in the more 'intuitive' response by journalists, pundits, academics, politicians and others who claimed the capacity to read the violence from the videotape. This symbolic violence took the form of what Lyotard termed 'democratic despotism', the judgement of the 'majority'.[5]

However, whether by means of democratic despotism or the terror of the sublime, to judge is to overwrite the integrity of that which is being judged, by means of another authority. In this sense, judgement is an infringement of integrity, and thus a form of violence. The madness of reason is that its grounds are arbitrary at least to the extent that they inform human judgement. This is simply the tragic predicament of modernity. To claim for oneself both origin and telos of history leaves one extremely vulnerable to the violence of arbitration.

We have seen this problem in previous chapters with reference to waste and risk. Like waste and risk, *violence is not*; but, as the videotape of the beating of Rodney King shows, *there is violence* (just as there is waste, and there is risk). To grant the intuitive more credit than merely that of prejudice, it is therefore better to work through an alternative which takes the *excess* (abjection) of violence as its point of departure. Violence is then not conceived of as integral to itself (an entity) but as a *differential* (a release of energy, for example as the splitting of atoms). As opposed to the integral which marks the attempt to capture the whole, the differential captures 'the essence' of transformation. This essence, however, is like the essence of a triangle (De Lauretis, 1989): a construct, a virtual object. This allows us to address and account for the intuitive anger that many people expressed in reaction to the verdict without reducing it to a mistaken identification of 'what really happened'. It also allows us to differentiate between judgement and intuition as distinct modalities of violence.

In order to rescue intuition from judgement, we must look more closely at the technological assemblage of media that were involved in the casting of the beating and its interpretations: . . . + *cars* + *bodies* + *batons* + *stunguns* + *handcuffs* + *camcorder* + *judicial discourse* + *photographs* + *television* + . . . (cf. Deleuze and Guattari, 1988). The differential of violence is not allocated to any medium in particular, but to the specific assemblage of *media hybridities*. This resonates Marshall McLuhan's enormously insightful comment that media hybridization is a release of energy in new forms in which violence is not an entity but an effect, a transgression. This is the *violence of mediation*:

> The hybrid of the meeting of two media is a moment of truth and revelation from which new form is born. For the parallel between two media holds us on the frontiers between forms that snap us out of the Narcissus-narcosis. The moment of the meeting of media is a moment of freedom and

release from the ordinary trance and numbness imposed by them on our senses.

(McLuhan, 1964: 63)

In short, media hybridities generate violence and different hybrid forms engender different forms of violence. It is not the nature of the media involved, but the specific assemblage in which they are being placed, that determines the specific modalities of violence that are actualized. Violence, operating in excess of any account that is rendered of it (logorationality), comes before judgment; it engenders judgment, because there is no judgment without mediation. Media hybridities disclose truth as an *event* and the truth of the event; this disclosure is violence. We thus have to look at the specific hybrid constellations to understand what forms of truth and modalities of violence have been engendered in the mediation of the beating of Rodney King and the verdict. The following three media hybridities have been central to the formation of the event that we associate with the beating of Rodney King, the acquittal and the subsequent disorders.

The first hybridity is between the *camcorder-videotape* and *legal discourse*. This hybridity is a very twisted one (Fiske, 1994). The ambiguity of the videotape, with its blurred images and lack of voice-over, does not dwell comfortably with the mode of rationality of legal discourse which is highly autopoietic. Legal discourse only recognizes its own constitutions as valid as a result of which its notions of truth and justice tend to be perceived as completely internal to its own discursive formations (Teubner, 1989; also see the discussion in Chapter 2). Legal discourse requires interpretations based on binary specificity: something is or is not. The lack of singularity of perspective of the videotape makes it difficult for legal procedures to enframe its meaning within such a binary logic. Three significant modifications have therefore been applied to the videotape in order to make possible its hybridization with legal discourse:

- The images were sharpened, enlarged, mapped onto a grid to specify different segments leading to a fragmentation of the process which was further intensified by various markings and highlights, such as that of Rodney King's tilted leg which – the defence attorney suggested – shows Mr King's threatening behaviour and his refusal to submit (Butler, 1993). A hyperreality was thus created in which more was to be revealed than could be seen by the untrained eye of the lay audience. A whole new dimension of reality could thus be revealed, ordered and manipulated.

- This kind of forensics addressed not only the intensity of visualization but also its nature. The fragmentation of the videotape was intensified by the use of still frames which transformed the *flow* of violence, the bodies acting upon another, the movement of the arm–baton–body vectors, into a *spectacle* of positions. The aesthetics of the spectacle are quite different from that of the flow. Whereas in the flow bodies are actively engaged in an event of which violence is the effect, in the spectacle bodies are frozen in postures which can only index 'intentions', i.e. force. Whereas the flow consists of conjugations, the spectacle consists of constellations. The flow is (be)coming. The spectacle is a *déja-fait* (Debord, 1994), indeed *passé*.

- Time is reversed. The time of the spectacle is a monad and embodies the smallest possible interval in which no motion is possible. By stopping the flow, the time of interpretation and reflection could subsequently be stretched almost infinitely. This allowed the defence to ask the officers questions such as 'what were you thinking at that very moment' while a still frame was being used to visually anchor that 'thinking' (granting it instantaneous (split-second) accuracy). The psycho-forensics induced by this temporal reversal enabled the officers to retrospectively insert their reflections into the event, as if they *were* thinking at the moment and thus conscious of their own actions. There is no contest. Reasonability was already inscribed into the event from the outset by the very media-hybridity through which it was formed.[6]

Out of this media-hybridity comes violence as an effect. The event of the trial was marked by a *symbolic violence*: that of acquittal.[7] This violence is marked by the imposition of a closure of meaning onto the sensibility of intuition, the meaning of which is completely arbitrary. The self-evidentness mobilized by this verdict, however, is internal to the medium hybrid in which it was cast. Outside that, as we will see below, *it did not make any sense*.

This is because most people were cast by a second media-hybridity. Instead of the camcorder–videotape–litigation hybrid, they encountered the . . . + *camcorder* + *videotape* + *broadcasting* + . . . hybrid. In contrast to the first one, this hybrid was very smooth. The ambiguity of the videotape was in fact extremely appropriate for the television medium (Fiske, 1994). Its blurred images, lack of voice-over and absence of editing intensified the sense of authenticity with which it emerged as an unmediated reality. This was not a staged event. Moreover, as the effectivity of television news is not based on presenting a single perspective, but always requires a doubling to turn an event into a contest, the absence of clarity and singularity helped the TV medium to take a step back. It did not have to provide a clear-cut judgement but could instead appeal to a higher truth consisting of two opposing viewpoints: guilty versus innocent. The meta-narrative of news coverage is the story that places two perspectives in opposition to another and allows itself to wither away from binary contestation (Allan, 1995; Cottle, 1993; Dayan and Katz, 1992). The telling of the story thus always grants itself a higher ground, a purer truth, even if this truth is only the inflated reflexivity that lies are being told.

The smooth merging of the videotape and the news broadcast realized a different form of violence than that of the acquittal. The violence it engendered was that of sensibility and mood: an *indexical violence* which does not resonate with the closure of meaning at the expense of alternatives, but with the affirmation of particular experiences and/or prejudices. For people who have (directly or indirectly) experienced the brutality of the LAPD, the beating itself came as no surprise, but there was perhaps a sense of anticipation that with the videotape their experiences would now be vindicated. For those (i.e. dwellers of the secured and regulated technological culture) who always believed that the police were there to protect and serve, the video-broadcasting hybrid might have engendered a violent disjunction of experiences in the form of disbelief. The 'stunning effects' of the videotape (and verdict), for example, induced liberal mediators into states of shock which could be rendered accountable only with the inoculation that the officers were 'rotten apples' (and the jury were misled). Others, less inclined to give up their own prejudices, might negate the intuitive truth of the images and resort to all sorts of denials, similar to those mobilized by the jury. In all these cases there is violence, positive or negative, that attaches itself to mood and produces sense on the basis of preconceived, hence particularly situated, sensibilities.

The third hybridity is a meta-hybridity between the first and the second. The trial was being broadcast itself, and in the news coverage of the trial, images of the beating were also appropriated to situate the verdict. For many people, the judgment of the verdict marked a complete denial of their intuitions and experiences; a denial of their sensibilities and sense of truth and justice (Hunt, 1997). This incompatibility within the medium hybrid produced a final and most violent clash between symbolic and indexical violence that in turn materialized into new embodied constellations that we associate with the 1992 'LA riots'. However, for those who already have direct or indirect experience with the colour of American justice, the verdict could have been attuned very well with their intuitions and in turn have amplified the sensibilities resonating with such intuitions. As Fiske (1994) and CBS reporter Ron Allen (*CBS Evening News*, 29 May 1992) both indicated, many African Americans already anticipated the verdict and were preparing for their response. The violence of these riots could thus be understood both in terms of outrage and euphoria as intuitions were both negated and affirmed.

Conclusion

With this case study of 'the Rodney King beating' the interpretative strength of conceptualizing violence as the effect of hybrid media has been illustrated. Different media-hybridities produce different forms of violence: symbolic, indexical and corporeal. The objective was to trace violence at the very grounding of racial identification. Racial difference, rather than constituting a lack with which violence can be identified, is itself already constituted by violent practices. As the case of the beating of Rodney King shows, these violent practices are not the exclusive domain of what Althusser (1971) termed 'repressive state apparatuses' but persist in judicial and political systems, as well as ideological apparatuses such as broadcasting media.

Whereas the concept of racism is generally invoked to explain the acquittal, Butler's (1993) and Fiske's (1994) accounts have shown that the imposition of a racial matrix both in the verdict and the subsequent riots was strongly mediated by different technologies of visualization, and signification, which in turn attributed different values to both types of violence. The technological dimension is crucial here because it reveals a sense of violence that is not contained by wilful bad faith or ideological distortion. It shows that racism does not need to be either consciously or ideologically articulated to be activated. The racial matrix through which violence against Black people was being justified was made operational not by a few racist ideologues and a gullible, ignorant mass, but by the very modern apparatuses of mediation, in this case of law and broadcasting.

By invoking the concept of risk, however, this chapter has added something to this crucial insight. It is not only that racism does not need ideological justification to work, but also its appeal to common sense is heavily endowed with a risk sensibility. As we have seen in previous chapters, risk is a highly volatile force. It enables a rapid and almost instantaneous transformation into opportunities as it flows between a range of different actors across multiple assemblages. Hence, the argument made by, for example, Gilroy (1993), Goldberg (1993), Miles (1989), Rattansi (1994) and Young (1994) that racism is an inherently modern phenomenon is strongly supported by the way in which race is associated with risk. These associations are made by technoscience, governance (policing, social services, public health, education, judicial apparatuses, law), media and commerce. However, unlike a reliance on either ideology or some mutation of psychoanalysis, the association between 'race' and risk enables us to see that the racial matrix deployed by modern institutions is not simply an ideological manipulation of self-securing identification, but structurally and technologically induced. The response to it from the allegedly 'gullible mass' is similar to the latter's response to many other perceived risks. In a securely regulated technological culture this response is more likely to endorse an instrumental fidelity and reliance on expertise. When technological culture crumbles, such instrumental fidelity discloses an ambivalence. It is here where new imperatives can emerge including those of other faiths.

In contrast to the violence of the acquittal, where technological culture still ruled (after all, this is how the system works), the violence of the rioting exposed an inherent vulnerability of this instrumental fidelity. Especially in the early stages after the verdict, people took to the streets an outrage that testifies of a refusal to be incorporated into the technological framing of the racial matrix. Indeed, technology broke down as a sense of risk proliferated in anxieties about miscarriages of justice, police brutality and socially endemic racism. The fact that media and government were able to regain a sense of control only after three days, and were only then able to provide a closure of the disorder, should not blind us to the immense impact of the aftermath of both the verdict and the rioting. It opened up much larger questions about social inclusion, risk, the role of media, as well as camcorders as technologies of countersurveillance (Thompson, 1995).

Violence as risk is never the property of a single medium, but is always the outcome of the meeting of two or more media. Violence as risk is an event of media hybridization. If we can agree that the mediation of violence is grounded in the violence of mediation, then this does not mean that we have given up on the responsibility to render an account of violent practices. All it means is that we attempt not to forget that every account that moves from intuition to judgement already entails a symbolic violence, a closure of sense into meaning at the expense of alternatives, setting into work a symbolic power by which particular symbolic associations are granted a claim to 'self-evidentness' and 'naturalness', that is, mythification (Barthes, 1957/1993). Instead of reasoning, we could perhaps better render an account of the madness of reason against the aesthetics of reasonability.

Violence as risk is the effect of hybrid media, the connectivity of technologies. Racialization is a particular form of violence, one that allows a closure of sense into meaning by which the *sense* of Blackness (e.g. the experience of slavery) becomes a *fact* of Blackness and subsequently the *risk* of Blackness; that which is received in silence. The silence of Rodney King, or that of O.J. Simpson, is in turn received by an exuberance of voyeurism and hermeneutic euphoria by media apparatuses which, instead of bearing witness, never stop placing meanings which perpetuate and extend the racial matrix formative of the prevailing structures of judging violence. The acquittal of the four police officers was exactly made possible because the sensibility of intuition, which is given in silence, was reconfigured in technologically mediated discourses which actively sought to constitute a reason within violence, a reason for violence, reasonable violence.

10 Conclusion

Risk and apocalypse culture

Recapitulation

It is now more than 15 years since the publication of Ulrich Beck's *Risikogesellschaft* in Germany. The question remains whether we are indeed witnessing a social transformation towards 'another modernity' predicated by risks. Beck sketched the contours of an emergent social (dis)order governed not as much by the principles of scarcity as by the principles of risk, including their avoidance and control. Indeed, towards the end of the second millennium, we were able to witness some (albeit modest) sustained changes in the organization of global capitalist society. In many western countries, calls have been made for new institutional forms and public bodies to enable the social and political reconfiguration of technoscience towards embracing a more responsive and publicly accountable logic of practice (Welsh, 2000). These calls often coincide with new political-economic forms that have gradually emerged from the waste left by two decades of monetarist and neo-liberal political economics, and take place under a plethora of labels such as, for example, the 'Third Way', 'Stakeholder Society' or 'Network Society'.

Although the world has been radically destabilized by the historic events of 11 September 2001, this has only further expanded the way in which risks are invoked to reconceptualize the modern social order. It seems now beyond doubt that our heightened sense of vulnerability induces particular risk sensibilities that enable a further spreading of a technological culture geared towards securing and regulating. Indeed, it is difficult to contest that we live in a risk society.

A point of contention, however, relates to the argument that risks are increasing both in terms of intensity and complexity. It could be argued that they are not, but that, instead, our ability to link different kinds of risks has increased. We have become more sensitive to the functioning of risks in enhancing global interconnectivity. Throughout this book we have stressed that such increased ability to articulate the complex nature of risk is inherent to modernity's technological culture. By extending Beck's thesis into the more Nietzschean and Heideggerian territories, occupied by Actor Network Theory and biophilosophy, we have seen that, far from antithetical to technological culture, risks are inherent to it. Hence, to call for more reasoned deliberation, institutional reform or increased reflexivity is futile as long as we continue to rely on reason alone.

To make this more clear we need to look at the nature of modern tech-noscience itself. Apart from colonizing the domain of 'beliefs' in terms of an instrumental rationality that is often directed against religious doctrine and mysticism, technoscience is also motivated by a desire for increased mastery over the future.[1] That is, it is never simply about revealing and ordering the causal aspects of phenomena, but also engages in revealing and ordering the specific purpose and effects of such phenomena, i.e. their futures. Technoscience seeks to intervene in the world and rearrange it according to certain modes of reasoning and programming.

Whereas Actor Network Theory is right to point out that such desires are only partial and to some extent practically counterproductive, the desire to overcome 'resistance' (reality) plays a central part in the subpolitical infrastruc-ture of scientific projects. It is on this basis that access to scarce resources is often legitimized and that scientists can create a sphere of autonomy by pro-jecting an expected date of delivery of the promises thus made. Even in the face of failure, technoscience can still maintain that these are temporary setbacks and that the overcoming of these obstacles is inevitable.

As Lyotard (1991) has argued, modern technoscience is thus motivated by an expansive desire for mastery and control over contingencies. This has rubbed off on more popular cultural sensibilities. Under the relentless driving force of technoscientific rationalization (which Weber (1985) referred to as the *Entzauberung der Welt*, the elimination of magic from the world), people have been invited to abandon the doxa of *fides*, and instead rely on an instrumental fidelity. There has been a slow but gradual demystification of everyday life throughout the modern era. However, whereas the mystical certainties of reli-gion may now have been displaced by the rational certainties of science, the shift of focus from God to 'man' has also engendered a different sense of suspi-cion, not over the question whether God exists or not, but whether we believe that what we now know is sufficient to secure our own destiny. The tragedy of modern technoscience has thus far been that its promises have not been fulfilled.

Latour's (1988) brilliant exposé of what certainly looks to be one of scien-ce's greatest successes, Pasteur's demonstration of germs causing diseases, is at the same time a story of quasi-accidental manipulation and inventive plotting. The highly dramatized public demonstration of the theatre of proof of how infections occur, however, also concealed the series of unknowns that pro-vided the links between causes and effects. It was only a few decades later that Martinus Beyerinck, in far less dramatic fashion, revealed a mode of infection that would – another half a century later – stain Pasteur's glorious victory over nature in the form of a virus, or in his terms, a *contagium vivo fluidum* (Van Loon, 2002). This shows a simple basic truth of modern technoscience – its success is always temporary. Nothing lasts forever. As the certainty of cause–effect predictions is always contingent on a *ceteris paribus* clause (that everything else stays the same), technoscientific assertions of truth always take place on the condition of underdetermination. Yet, at the same time, this contingency of truth has to be denied. In the long run, all contingencies will be accounted for.

In a culture dominated by technoscience, i.e. a technological culture, contingency is risk. It is that which has not yet been charted; that which is manifest only as blanks on a map, black boxes in a flow chart, silence in a discourse. That is, rather than complete absence, which would prevent risk from being perceived at all, risk is the presence of absence: something is there but not (yet) fully revealed. The tragedy of modern technoscience is that while it seeks to expand its control over contingencies, it produces more contingencies (Pacey, 1983; Winner, 1977). What starts as little deviations and anomalies may always have the potential to completely undermine the basic assumptions that were taken for granted, the facts that were considered to be indisputable. The only way in which technoscience can continue its quest for both completeness of understanding and prediction is by bracketing off anomalies, postponing the rendering of an account of them until the general picture is complete (Lakatos, 1978). Needless to say, during this 'meanwhile' suspicions are likely to increase.

Such negative heuristics would still be manageable if technoscience only operated in a vacuum. However, the very nature of the technoscientific quest is that it actively intervenes in its environment and modifies it accordingly. It also affects the environments of other systems of social ordering such as governance, law and commerce. This is why risks engendered by technoscience constantly 'spill over' and flow into other social institutions (Feenberg, 1991). This is why technological culture activates risk sensibilities. This is also why, despite increased efforts towards securing and regulating contingencies (which prompts the suggestion that we are now safer than ever), risk sensibilities have only intensified.

Throughout this book we have seen examples of risks as markers of the limits of technological culture. Waste, pathogen virulence, cyberrisks and collective violence are but four examples of myriad 'bads' that have affected the organization of modern society. More specifically, we have seen that:

- the technocultural enframing of risks transforms risks into opportunities and thereby engenders more risks
- such risks are endemic to the flows that are part of the intersystemic constellation of visualization, signification and valorization
- the main response of the technocultural system is to cultivate a moral panic and institutionalize a culture of risk aversion and thus risk containment
- the technologies of mediating the organization of the risk-aversion society are always on the verge of breakdown because the intersystemic constellation in which they are set to operate cannot be permanently integrated.

In other words, the sense-making of risks and cultivation of risk sensibilities are themselves forms of social and symbolic violence that are processed back into the technocultural system as reinforcements. These four elements constitute the notion of virulent modernity. The elements that are identified as 'foreign bodies' (i.e. risks such as waste, pathogens, cybercrimes (including addictions) and violent disorder) cannot be eliminated, because they are endemic to the very logic of organization of modern social systems.

Hence, instead of an increased sense of security, the very technological systems geared towards securing and regulating have engendered increased risk sensibilities. Nowadays it is not uncommon for people to feel overwhelmed by the sheer amount of information about risks in their everyday lives. This may in some cases create a sense of defeatism: information overload makes rational responses to risk practically impossible. Combined with a growing force of individualization, in which more and more essential decisions are thrown back onto individual responsibility, it becomes difficult to develop rational strategies for dealing with risks. If everything is risky, then risk loses its appeal to sensibility. In the face of an ever more elaborate system of information provision about risks associated with food, for example, there is a threshold beyond which there is no rational basis for a determined reflective response. In such cases fatalism and apathy will prevail, or perhaps a renewed sense of 'fides' attuned to a non-arbitrary divine power.

The theoretical framework from which these insights have been derived consist of three main approaches: risk society thesis, Actor Network Theory and biophilosophy. Although they are not always complimentary, these three perspectives share an alertness towards ambivalence and the crucial role of anticipation. We have used Heidegger's reflections on technology as enframing/revealing to conceptualize technological culture as the common ground of all three. His suggestion, that modern technology has turned nature into a resource, standing reserve for whatever use 'we' may have in mind, is particularly adept in clarifying the devastating ecological implications of so many of our modern technologies.

We only need to look at the phenomenal problems caused by 'waste' and the pressures exerted by waste disposal on the environment to understand the enormity of the impact of modern technology (Cutter, 1993). However, at the same time, we fully depend on the same technological culture to provide solutions. The reduction of waste is primarily seen as a technological objective; for example, the notion of recyclability is now incorporated into the design of many products, and new technologies are constantly emerging to deal with the recuperation of waste. This double-bind is what Beck (1995) has identified as the crux of the risk society: industrial modernity has evolved through the spreading of technoscience and its applications to all domains of human life. Technoscience offers its members (users) a specifically powerful mode of enframing/revealing. It orders the world on the basis of a methodical rationality. 'Facts' are established through highly regulated procedures which at one and the same time enclose a particular universe of possible statements and excommunicate any modality of expression that does not conform to the assumed-as-given procedures. Hence, the modality of discursive formation offered by technoscience incorporates politics and religion, as well as common-sense cultures, and seeks to subdue these to its own level of engagement.

Beck's understanding of technoscience comes close to Weber's (1956) notion of *Zweckrationalität* (literally meaning 'goal-rationality') which is the principle that governs bureaucratic organizations. It refers to an operational logic of cause-and-effect means-to-an-end thinking. The principal aim is to identify the means by which to obtain certain previously established ends. Students of bureaucratic organizations have subsequently pointed out, as did Weber, that after a while means become ends in themselves. For Beck (1997) this is also a key aspect of the risk society, in which the instrumental rationality of bureaucracy, a characteristic feature of most organizational settings of modern institutions, drives a vicious circle between risk and risk management, and between complexity and ambivalence (undecidability).

Although Latour (1993) made a considerable effort to dispense with the idea that there is anything specific about 'modernity', and questions its adequacy for understanding technoscientific practices, this critique has not found many followers among contemporary social and cultural theorists, most of whom have implicitly embraced Foucault's analysis of the episteme of modern thought as discussed in his *Order of Things* (1970). That is to say, very few theorists have actually been 'modern' according to the caricature offered by Latour (as linear-thinking, rationalist programme-oriented technocratic engineers) and, like Weber, never dismissed the implicit ambivalence of instrumental rationality which can be easily exposed when pursuing its most radical forms (e.g. the madness of reason of bureaucratism).

However, another aspect of Latour's critique on modern social theory such as Beck's remains strong: that it fails to consider a distinction between technoscientific epistemology (science) and its messy, sometimes confused, but above all political and pragmatic social practices. This is not to say that the basis of Latour's critique of social theory is the latter's lack of empirical work. Of course, this would have been a valid criticism as well, but Latour's point is much more fundamental. Social theorists ignore the empirical reality of technoscience (or any other practice) because it has separated 'reason' from 'reality' (or force). In so doing, it confuses the strategy with the actual work being done, and thus reduces the unfolding of history to a narrative of programming and engineering. Latour wants to bring back an element of contingency or 'accident' into history; that is to say, by focusing on the specific arrangement of actor networks that enframe/reveal particular scientific facts as reality, he dissolves the artificial distinction between the representational and the actual, and thereby also breaks with the narrative continuity of modernization as progress without having to resort to a postmodern apocalyptic narrative of pandemonium and 'ends'.

In the discussions of waste and emergent pathogen virulence, the strong modernization-focused risk society thesis could be defended only at the expense of omitting questions of *translation*. For example, the question of how technoscientific solutions of recycling in conceptions of waste management translate into an actualization of waste-handling cannot be addressed because the latter are primarily conceived as 'means'. The relative autonomy of the technologies of visualization, signification and valorization is not really appreciated as waste management is reduced to strategies and programmes. The fact that the handling of waste has become a massive economy as such does, however, illustrate Beck's point that risks may always become opportunities and serve in the reordering of the economic infrastructure of modern society.

With pathogen virulence, the risk society thesis is not as easily recuperated. This is because pathogens such as viruses and bacteria are far more easily understood as actants than waste. They act and react intelligently, and are able to adjust their strategies to changes in their environment, apparently over a relatively short time span. We need a far more detailed scrutinization of the technologies of objectification of pathogens, and how they translate into clinical practices. Moreover, we need to be able to engage with the micrologics of parasitism that underscore the motivation of these pathogens. What this means more concretely is to understand interactions between pathogens and humans as a multiplicity of forces. Pathogens are never simply agents of diseases (and careers); they are also, for example, powerful political allies, fascinating personas, blockbusting movie stars, cunning litigators and smart military strategists. Each of these different roles could be taken by the pathogen-actor at the same time, depending only on the network-constellation into which its being is inaugurated. The problem with the risk society thesis is that it is too human-centred and its ethos is too humanistic. There is an implicit assumption that if we would deliberate properly, in an environment free from irrational interests and short-sighted power play, we may find a proper mode of reasoning about and with risks. Beck already knows that this is not going to happen. A key factor here is time. Proper reasoning, proper reflection, proper representation and participation all take time. Risks, however, constantly foreshorten the future; they instil an urge to act now, immediately. The risk society is immersed in the politics of urgency; the logocentric technological culture upon which it thrives cannot cope with the speed and versatility of risks; they constantly engender an ethos of emergency and haste.

Emergent pathogen virulence is increasing. It correlates with other ecological problems such as deforestation, climate change, and the pollution of air, soil and water, as well as social ones such as increased warfare and collective violence, the growth of the human population, malnutrition and increased industrial productivity and waste. Moreover, it interacts with a technological culture that is becoming increasingly apocalyptic, that is, revelatory. The risk society thesis too is part of this technocultural apocalypse, in the sense that it generates reflections that do not offer an escape from ambivalence (e.g. through creating a false hope in, for example, certain modern institutions, policies or political procedures). However, it is not dystopian. That is, ambivalence is not 'an end', it is merely a temporary suspense of logocentric ordering, of a metaphysics of presence that mythically closes the gap between the virtual and the actual by the sheer force of will.

It is this ambivalence that also marks the work of Latour, who, although far less apocalyptic, also refuses to bow to the metaphysics of presence of logo-centric ordering; at least to the degree that there could ever be an equivalence between strategy (reason) and practice (force). However, unlike Beck, Latour does not reserve a special role for the pathological (e.g. risks) in his under-standing of actor networks. Actor networks are formed around a multitude of interests, among which risks. The problem perhaps here is that, for all the positivity of force, the question remains why actants would want to seek to enhance their strength by connecting to others. Self-preservation seems one of the few motivations that could apply to all kinds of actor networks as described by ANT. Risks, of course, also signal the ontological problem of 'preserving' the self, although by this every definition, self-preservation becomes primarily marked in terms of status quo, that is, the interests of the state apparatus.

Although ANT fills significant gaps in the risk society thesis, it suffers from similar problems regarding the question of motivation. Both RST and ANT place too much emphasis on interests, on self-preservation, on enhancing strength, in short, they are too much concerned with the human logic of ordering according to individual motives. They have unduly singled out *auto-poiesis* as the vital force of modern being. As a result, the main response against risk can only be conceived in terms of 'immunization' – preserving systemic integrity of actor networks or institutions against outside interference.

Against the autopoiesis of risk aversion: symbiosis and the nomadic war machine

It is in biophilosophy that we can find an alternative to this politics of immu-nization. The question that both RST and ANT cannot address is: what moti-vates a virus? For most people who have been brought up in Darwinist culture, such as Richard Dawkins, this is an easy question requiring a simple answer – the reproduction (survival) of (selfish) genes. Like Thatcher's market (non-) society from which it emerged, Dawkins's cosmos consists of purely selfish creatures. In such an environment, evil means nothing because there is no free will. Pathogens have to destroy in order to survive, the choice of sacrifice is not there. However, anyone familiar with symbiosis theory knows that such an answer, for all its assumed cruel elegance, is simply insufficient because, if anything, the evolution of viruses shows that they often mutate or, worse for selfish genes, acquire a skill of association with a host, by means of which both gene-pools lose their distinctiveness and become entangled in a synthesis of information flows. This symbiosis, even if motivated by selfish genes, is any-thing but selfish because the 'self' ceases to be and becomes other. Hence, even if we were to follow Dawkins's evolutionary biological line, we would still end up in a strange place, where unintended outcomes prevail over intended ones. From this we could deduce that what motivates the pathogen is evil if it is genetic selfishness. However, by the same token, its motivation could also be to sacrifice itself to enter into a higher state of symbiosis, i.e. communion with the other. Hence, we have two kinds of pathogen motivation: one is self-replication, or autopoiesis, the other is transcendence or symbiosis. Whereas it would be premature to suggest that viruses (risks) have a free will or that they can chose between being autopoietic or symbiotic, one could argue that there are at least grounds for introducing this fundamental choice at the heart of 'parasite politics' (Van Loon, 2000b).

The essence of biophilosophy, it could be argued, is that it engages difference as difference. The concept of symbiosis plays an essential part in providing an insight into understanding the formation of assemblages. Indeed, symbiotic assemblages could be seen as a form of parasite politics. Ryan's (1996) concept of junctural zone (Chapter 6) as the space in which two different species meet, a liminal space where difference matters, has enabled us to understand the less-than-smooth operations of assemblage formation as sites of violence, risk and anxiety, indeed political struggle.

The concept of assemblage points towards the centrality of intersection, exchange, transgression, transformation, multiplicity and above all difference. It suggests that our commonly taken-for-granted notion of 'integrity', whether it is applied to systems or 'individual' entities, is highly problematic. It suggests that singularity, or the principle of irreducibility as Latour (1988) calls it, is not the same as individual integrity. That is to say, singularity does not correspond with 'autonomous wholeness'.

The desire for autonomous wholeness (the Oedipus Complex), however, is deeply inscribed in the logic of modern being. It constitutes the whole domain of psychoanalysis and has infiltrated our culture in its deepest core (Deleuze and Guattari, 1983). This is why evolutionary biology often seems to justify utilitarian selfishness: the will to survive corresponds so well with the will to self-preservation. This logic constitutes both market economics and representational democracy. Producers and political parties supply 'goods' for consumers and electorate, whose decisions merely follow calculations geared towards maximizing self-interests.

Indeed, modern technological culture seems to thrive on a cult of individualism, selfishness, greed, and – as a result – carelessness towards others, most vividly represented in carelessness towards one's environment. However, the same technological culture has also given birth to welfare state provisions, systems of care, collectively organized forms of altruism, etc. Of course, one could argue that these provisions are themselves merely effects of self-interested actions. This argument is championed by defendants of rational choice theory, for whom 'good' inherently corresponds with maximizing benefits.

But we need not stoop that low in terms of denying people genuine motivations and free will (to act unselfishly, if they wish). We could also argue that most 'other-interested' initiatives of modern society actually stem from discrepancies in the modernization process itself. It has often been argued that the welfare state was not formed as much by liberal or even socialist ideologies, but by forms of charity that have a distinctively non-modern origin in religion or even pagan (tribal) ethics (De Regt, 1984; De Swaan, 1989) based on 'agnatic solidarity' (Deleuze and Guattari, 1988). Finally, we could argue that modern technological culture has never been as consistent as Heideggerian or utilitarian theory has suggested. That is to say, as Berman (1982) and Bauman (1990) have argued, modernity is itself an expression of ambivalence.

Focusing on the general phenomenon of risk, we could argue that it reveals an irreducible ambivalence of technological culture. Risk marks a moment where technological ordering 'breaks down' – it discloses that technology is revealing. Risk forces us not to take the taken-for-granted for granted. Bernstein's (1996) argument that what characterizes modern culture above all is the ascendance of 'man's' control over risk – while confusing a desire for control with its actualization – signals that modernity has always co-evolved with risk. However, regardless of whether he is right to give it such a prominent position, his own analyses of risk-management strategies in stock markets show that there are never risk-free opportunities. That is, whereas 'man' has cultivated a habitus of controlling risks and maximizing opportunities, risks always escape such enclosures. This excess nature of risk manifests itself in a range of virulent abjects, of which we have discussed only a few.

The ambivalence of modernity is that it is never purely autopoietic. By desiring 'progress' and 'transgression', it inevitably finds itself in need of 'foreign' codes to translate new systemic operations of newly colonized territories. Entropy destabilizes. Nowhere is this more clear than in the history of risk, which, starting with a narrative of regularity and security, has now become a story of irregularity and insecurity. The more contingencies it incorporates, the more its desire for immunity (integrity) becomes counterproductive. This pattern manifests itself, for example, in various financial and legitimation crises of governmentality and the (nation, welfare) state.

By contrasting the state apparatus with the nomadic war machine, Deleuze and Guattari (1988) suggest that throughout history there have been two distinct ways of ordering social systems. The state apparatus is geared towards securing and regulating space in terms of specific categories, enframed by distinct boundaries. It works on the basis of territorialization. The nomadic war machine is geared towards movement, dissolution, creating an open, smooth space of deterritorialization. These two forces of ordering are also prevalent in our technological culture. The first reflects the established institutions of modern society: technoscience, governance, mediascape, commerce and especially law and the military. The second refers to more marginal forces, that sometimes operate within and sometimes without those institutional boundaries.

The argument in this book is that risk can be seen as an opening up of the state apparatus, or more specifically, risks have enabled nomadic war machines to escape from the clutches of the logic of disciplinary power. However, this has by no means been a one-dimensional movement. During its inception, the modern state apparatus has itself appropriated risks in support of the struggle of 'man' against nature, divinity and history. Risk has been wrestled out of the hands of the gods, Bernstein (1996) has argued, so that 'man' can start to control contingencies and 'his' own destiny.

Indeed, even today we can clearly see how risks have proven to be an ally of the modern state apparatus. Risks have been called upon to increase the allocation of resources to technoscience, to further governmental control over social processes, to create concerns and spectacles for mass mediation, to induce more needs for consumption and thus market opportunities for commodification, to expand law and, last but not least, to enhance the perceived need for military command, control and intervention. Indeed, in as far as it is concerned with risk aversion, the risk society is not at all another modernity, but simply a more intense version of the same social formation from which the same risks have emerged.

The science of risk assessment and risk management is thoroughly modern, even when it embraces 'complexity theory' to incorporate undecidability and unpredictability. For as long as there is a belief in the rational calculability of contingencies, there is hope for their ultimate eradication with the further expansion of telematics. However, at the same time non-linear thinking undermines a faith in such increasingly complex calculations. That is to say, whereas fractals for example do not mean that 'anything goes', the fact that one can arrive at such highly complex, underdetermined forms on the basis of an essentially simple formula arouses a suspicion that as the complexity of such formulas increases, so will the underdeterminability. The question then becomes, at what stage do we cease to think in terms of underdetermination and start to accept *undetermination?*

This 'breaking point' is the threshold, or turning, of technological culture. Whereas risks have been instrumental in the advance of the technocultural systems of science, governance, mediation, commerce, law and the military, they have also engendered their gradual erosion. Indeed, modernity has always been a force of entropy – as it spirals outwards to colonize more and more aspects of life and encodes them in the *logorationality* of instrumental fidelity – it also slowly but gradually disintegrated, first at the margins but steadily also towards the centre. Other, foreign, aspects have cropped up at the very heart of logorationality, exposing the madness of reason in the form of completely irrational assertions to have faith in science, governance, markets or the law. The acquittal of the four LAPD officers discussed in Chapter 9 is a clear example of this madness. The cracks that are the consequence of this entropy are starting to show, creating more anxiety and fear over risks, inducing more risk aversion, further undermining the stability of technological culture.

Apocalypse cultures

In 'The Question Concerning Technology' Heidegger argues that revealing must be seen in its widest possible sense. Not only in terms of 'making visible', but above all in terms of 'ordering' (classifying and commanding). Heidegger asserts that technology is thus not a tool, although in our practical ordinary everydayness, we tend to reduce technology to tools. This generates the instrumentalist ethos with which modern 'man' has rearranged his natural environment according to 'his' own needs, often regardless of the wider ecological issues of living. Although he did not use the term as such, technological culture could be seen in Heidegger's work as synonymous with modernity. It is the casual and careless inhabitation of an environment of artefacts, 'ordered' and revealed by technology, that defers any contemplation or reflection on itself unless things 'break down'.

Risk, of course, is an anticipation of such a breakdown. Hence, risk signifies the limits of technological culture. It is through risks that the essence of technology as revealing becomes highly visible in a very practical sense. It is important to stress here that what is being revealed is also our careless inhabitation of technology. We have been using mobile phones for years and in good faith before some concerns were raised about possible negative side-effects such as brain tumours. The same goes for British beef, genetically modified (GM) foods, measles, mumps and rubella (MMR) vaccinations, etc. The risk society thesis suggests that we may have arrived at a point where we are no longer able to carelessly inhabit technologies, because we now anticipate risks everywhere, regardless of the increasing investments in safety technology and security concepts. Does this mean that we are arriving at a point where technology becomes more revealing than ordering, more unconcealment than concealment? The world of cyberrisks seems to suggest that this is indeed becoming the case.

The Italian philosopher Gianni Vattimo (1992: 15) has taken up this notion of this unconcealment with reference to telematics. Vattimo argues that telematics has intensified the revealing-ordering ability of modern technology in terms of a rendering visible of everything at immense speed. However, far from bringing greater insight into the order of things, telematics has the paradoxical effect of eroding the principle of reality (for example, hyperreality).[2] It implies the reduction of the world to images. Essential to the image is its reduction of history (sequence, causation) to the level of simultaneity (Barthes, 1977). Overexposure and immediacy have generated a world in which performativity and signification become indistinguishable (cf. Virilio, 1986). As we have seen with reference to 'Nature' or 'the' environment (see Chapter 8), the intensity and speed of telematic symbiosis have thus undermined the very foundation of 'authenticity' as 'truth' and obliterated the distinction between the 'proper' and the 'fraudulent'. As a result, 'fraud' becomes endemic and is indistinguishable from 'truth'. It is not surprising that 'cybercrime' plays such an essential part in the telematic revolution, and has actually generated most innovations in this new world information order.

This has profoundly disturbing effects on the very epistemic formation of modern thought, which as we know from Foucault (1970) is based on the principle of linear causality (also see Adam, 1990, 1995). It has led to 'a kind of entropy linked to the very proliferation of the centres of history, that is, of the places where information is gathered, unified and transmitted' (Vattimo, 1992: 22). The singularity of perspective, which was ordered and revealed by modern technology and constitutive of the humanist ideal of self-transparency, is shattered by the visual cultures of the new ICTs.

> Instead of moving towards self-transparency, the society of the human sciences and generalized communication has moved towards what could, in general, be called the 'fabling of the world'. The images of the world we receive from the media and the human sciences, albeit on different levels, are not simply different interpretations of a 'reality' that is 'given' regardless, but rather constitute the very objectivity of the world.
>
> (Vattimo, 1992: 24–5)

Referring to Nietzsche's aphorism of truth as a fable, Vattimo suggests that our world is ordered through information and communication flows that evolve around the production and dissemination of a range of 'fables' about who, what or how we are, why we are here and where we are going. More and more of our world takes the form of fables, of *assertions* of truths (performativity), of facts, of stories, of symbolic relationships and orderings, that have no external reference point but their own constitutive myths.

In other words, we are facing a paradox of two simultaneous movements. On the one hand, the accumulation of risk perceptions undermines technological culture as an ordinary-everyday pragmatic appropriation of tools. On the other hand, the proliferation of telematics has generated a fraudulent illusion of transparency in which a multiplicity of coexisting myths continue to order and reveal a world that is becoming more and more hyperreal.

Hence, whereas we can agree with Heidegger that the essence of technology is revealing, we must also acknowledge that technology is far less effective than he seems to suggest. Especially in the modern era, this revealing has taken the form of ordering what is being revealed into a particular constellation. Although ordering and revealing always come together, it is only in our modern (telecommunication-based and mass-mediated) technological culture that the ordering turns everything into a standing reserve as datafiles, ready to hand for processing whenever it is deemed necessary (Poster, 1995).

The limits of this telematic transformation are clearly exposed by cyberrisks, in particular internet viruses and hoaxes, which operate alongside battles between encrypting and hacking, circumventing security systems, obtaining confidential information, defrauding individuals and institutions, engendering addictions, paranoia, anxiety, blackmail, etc. Cyberrisks transform autopoietic systems into symbiotic systems. This is the essence of cybernetics: the formation a totalizing, all-embracing monad.

The crucial turning of technological culture is that en-presenting increasingly entails a 'fabling of the world', transforming a relatively stable (albeit highly dynamic), symbolically ordered 'reality' (e.g. 'modernity') into a fragmented hyperreality of empty signs: signifiers without signifieds. That is, the paradox of technological culture as revealing its own forgetting yet continuing its fabling of the world manifests itself in discontinuities between visualization, signification and valorization. Whereas we are often able to anticipate risks in general, we find it difficult to attribute meaning to anything specific, hence it does not enter either political or economic processes because a general risk-anticipation has no value. The selectivity of what matters and is deemed 'worthy of concern' is arbitrary yet inevitable. Moreover, even if we agree that some risks matter, we find it often impossible to ascertain how much, which inhibits systems of governance and commerce to organize any concerted effort of dealing with or capitalizing on such concerns.

The concept of en-presenting thus enables us to understand why risk perceptions may be increasing yet fail to have any impact on how we conduct our affairs. Hence, it is not that we are totally blind to our own blindness (most of us know that we do not see properly), but we find it simply impossible to attune our vision, language and taxonomy to the immensity of the tasks we are facing. Haraway's (1988) concept of 'partial knowledge' is a particularly useful way of conceptualizing this problem. The assemblages through which we work towards becoming-human – the principle of being modern – cannot maintain consistency. The monad turns out to be not 'man' but a machine, an information-processing system that seeks to produce thought without a body (Lyotard, 1991). This is the biophilosophy of apocalypse culture. It enthusiastically borrows from Nietzsche attempts at a new conceptualization of being, i.e. 'viroid life' (Ansell Pearson, 1997). It dwells on 'the turning' that fails to be completed. It articulates an ethos of the interregnum, a hegemonic disequilibrium in which the old order is no longer valid but the new cannot (yet) be born.

It could now be argued that the concept of risk embraces both sides of the turning. When its use initially became widespread, mainly through applied mathematics in, for example, insurance industries and stock markets (Bernstein, 1996; Boden, 2000) as well as epidemiology (Anderson, 1983; Lilienfeld and Stolley, 1994), it was used as a means for the extension into the future of technological culture's drive towards an increased regulating and securing of contingencies. Hence, 'technologies of risk', i.e. calculations of probabilities that something bad might happen, including the extent of this badness, were primarily used to order and reveal nature, divinity and history and to subject them to 'man's' will to power. In this sense, risk was not at all operating as a turning. The dangers revealed by technologies of risk were simply the saving power of technology leading to more regulating and securing, unlocking and stockpiling of resources. Technologies of risk thus conceived merely engendered more structurally induced needs, varying from an ever growing cocktail of vaccinations and ruthless antibiotics to virus protection software and seatbelts. Propelled by specifically cultivated risk sensibilities, they are the former additional accessories of modernizing systems that we have become completely dependent upon for our continued successful engagement with capitalist technological culture.

Indeed, for a long time, the language of risk was identical with that of risk aversion. However, Beck's risk society thesis suggests a more insidious and unsettling movement alongside the extending of the present by risk anticipations. Whereas the concept of risk has itself become 'endemic' in our technological culture, we have also experienced the 'downside' of our increased dependence on structurally induced needs. Catastrophes such as Chernobyl and Bhopal, for example, have shown us what can go wrong, even in so-called 'high-security' risk environments. Paired with technological breakdown, risk anticipations are now beginning to turn against a culture of complacent 'incorporation'. They have now been extended to the very technologies of risk that were devised to increase regulation and security. The technoscience of risk is suffering from delegitimation. That is, rather than increased security, they now also contribute to increased suspicion, doubt and scepticism, and no longer just among hippies. For example, today many mothers have become experts in domestic 'safety issues'; the increased market share of organic products testifies to this, as does the expansion of homeopathy. Worldwide protests against GM foods and globalization – as well as what are often (and erroneously) considered to be right-wing 'life' issues such as abortion – also signal that the opponents of modernity are not a mere lunatic fringe of people caught in a time warp.

Whereas many issues are still dominated by a culture of risk aversion, the revealing of structurally imposed needs also creates a growing sense of discontent. This signals a 'turning' within modern technological culture. The dangers which were the saving power now start to overshadow the promises of eternal bliss at the end of history. Instead, our technological culture has become more apocalyptic. As technological culture merges with apocalypse culture, the social, political and economic spheres become more volatile.

Whereas technological culture moves us to embrace the future with a firm and certain grasp, apocalypse culture reveals that we are clutching onto straws and that we are approaching the end of 'our' civilization. Whether or not we actually believe that western modernity is to be the first 'civilization' that is going to last forever – all historical evidence points towards the opposite – it is clear that at this stage it still rules supreme in the world, even if it is showing some cracks and fissures. However, the apocalypse culture that accompanies the risk society does undermine the relationship between concealment and unconcealment that is at the heart of technoscience. For example, by highlighting that the human being is merely 'a face drawn in the sand at the edge of the sea' (Foucault, 1970: 387), 'man's' assertion of domination over nature, divinity and history is also questioned. This romantic backlash against technoscientific reason is of course well known and is – for example – well illustrated in Nietzsche's concept of the Übermensch, the transhistorical figure that is to replace the human being as the heroic subject of history.

Apocalypse culture is of course nothing new. Indeed, parts of it are rather transhistorical apparitions. It is important to note that it comes in two – oppositional – forces. On the one hand, apocalypse culture is engendered from within. This immanent apocalypse culture manifests itself in terms of a subversion of all moral values, including the most sacred taboos, of the dominant civilization from which it emerges. Adam Parfray's (1990) *Apocalypse Culture* contains some excellent examples of this. It is a collection of writings on the most extreme perversions and subversions such as necrophilia and automutilation. In this sense, immanent apocalypse culture is marked by a hedonism illustrated by the phrase *carpe diem* (seize the day). All values are to be destroyed for they are all hypocritical assertions of the dominant culture's will to power. This 'abasement' of cultural values is often confused with the decadence that is associated with it. The only – but essential – difference is that decadence takes place within a dominant culture and does not in itself question the hierarchy of values. Indeed, it is most effective and prolific if these values are – even if perversely – upheld. Decadence is the full realization of hypocrisy. Immanent apocalypse culture goes one step further – it exposes it and celebrates its collapse. Hypocrisy is its enemy.

This is very different from the transcendent apocalypse culture, which is a transhistorical force. Here, the main impetus for undermining is inspired by 'a beyond' of the civilization whose end it proclaims. It manifests itself most strongly in opposition to both the immanent apocalypse culture and the decadence of the dominant culture by asserting that the end of the particular civilization is caused by its moral disintegration. Transcendent apocalypse cultures are usually religious; they are often associated with cults and sects, but sometimes with churches and even political movements. They speak of the wrath of God (or Nature, sometimes even History, i.e. militant Marxism), and the moral decay of civilization could be seen as caused by human beings' fatal conceit that they represent history, nature and/or divine power.[3] The narrative of the Fall repeats itself throughout history, even in non-religious explanations such as Nature hitting back.

Combining the writings of Beck, Latour, Haraway, Deleuze and Guattari, Lyotard and Heidegger with the four illustrations of Part II, we can conclude that the risk society has given rise to an emergent apocalypse culture, which manifests itself both in its immanent and transcendent forms. That is, the turning in how risks are being perceived has undermined technological culture's ability to enframe everything as a standing reserve. Instead, visualization, signification and valorization have become increasingly disjunct, with incompatible en-presentings of risks. What we need to consider is how technological culture deals with this emergent apocalypse culture, both in its immanent and transcendent forms.

Fidelity

The birth of the risk society has been an inconspicuous and long-term event. It started with the earliest assertions of 'man's' systemic dominion over nature, divinity and history and accompanied 'him' along his trajectory of self-valorization. In this sense, risk is endemic in technological culture. However, as we have argued, a crucial break occurs when technological culture can no longer perform the closure of ordering; i.e. when it can no longer promise to rule out the immanence of irregularity and insecurity. Perhaps the most crucial sign of this new disclosure of technocultural fallibility has been the delegitimation of science. It is finally beginning to dawn on governments, businesses and technoscience that people can no longer be approached through a simple propaganda model of using media to convince the public that certain processes are safe whereas others are risky. This is because a basic sense of 'trust' between people and modern institutions (subpolitics) has all but vanished from the public sphere.

A change in 'trust' is the crux of this transition. In our modern, western tradition, trust is a dense and oversignified term that designates at once a property (one 'has' trust) and attribute (one trusts someone else if that person is 'trustworthy'). It is thus something one 'has' and 'invests' (thus 'gives'). Following this convention, trust is an entity that can be attached to a value attributed to someone or something other; it is a crucial part of valorization and thus of the cycle of credit referred to by Latour and Woolgar (1979). This is the paradox of trust: one can have trust only if it is given/attributed to someone else. Trust always entails a particular (even if generalized) 'other'. 'I trust' makes no sense, 'I trust you' does.

This paradox becomes less taunting if we accept the proposition that trust is not an entity but a force that modulates (attunes) relationships. In this sense, trust qualifies and valorizes a particular form of social interaction and bonding. What the qualification of trust adds to a relationship can be a range of attributes: confidence, reliability, commitment as well as security and endurance. Building up trust is generally speaking a time-consuming process, but once established it tends to endure and prolong itself. Trust connotes a stability of the relationship in time; it is a temporal form that outlasts the present. It is this sense of trust, as attuning social interaction, that allows us to link trust and sociality, and more specifically to theorize trust as a *necessary* condition for sociality.

Trust is relational. The specific modality of this relationship, however, is less clear. Giddens (1991), for example, defines trust in terms of reliability. To invest trust in a machine, institution or expert system is to attribute a positive evaluation to its reliability. As has been argued in Chapter 2, for Giddens this trust/reliability nexus is at the heart of modernity. Modernization is inherently a process of detraditionalization (also see Heelas *et al.*, 1996). This detraditionalization has led to the erosion of 'doxa' – systems of belief, materialized in rituals, institutions, habits and practices, that provide a sense of ontological security about the place of the human being in the cosmological order of things. Secularization is often seen as a central force in detraditionalization, but so is the emergence of science, the mass media, modern art and more recently perhaps 'postmodernism', which is inherently related to the collapse of all foundations and (moral, ethical, logical) securities upon which modern society relies, including, truth (science), justice (law, politics) and aesthetics. In other words, modernization generates an innate sense of ontological insecurity about who we are and how we should live.

With the rise of ontological insecurity, Giddens (1991) observes a parallel emergence of new institutions and expert systems, often embracing the forms of bureaucratic rationality that Weber described more than 80 years ago (Sayer, 1991). The main rationale for these institutions and systems is to provide new ontological securities, in the form of scientific truths and legal-rational forms of justice. Indeed, there is very little in the modern world that is not somehow connected to scientific or judicial-political practices. However, the main difference with the traditional doxa for Giddens is that these institutions and expert systems operate on the basis of heterodoxy. This means that there is no longer a single path or truth of justice but multiple ones, often existing in conflict with each other. For example, in recent ecological or health-related issues (such as the BSE/vCJD scare in Britain, see Ford, 1996), scientific, political and legal arguments often operate on totally different grounds, leading to a highly complex web of discursive assertions and debates, which are often confusing and misleading. We moderns trust expert systems to be reliable because we have no other means of assessing and evaluating their adequacy.

It should be clear that there is no reciprocity at work here. 'The public' can never be in a relationship of equality with the experts; they do not have the same sort of access to information nor do they have the same capacity to decode that information or translate it into public action. This is quite clear, for example, in the case of environmental activism, where experts working for non-governmental organizations often take a leading role in raising issues and concerns (Macnaghten and Urry, 1998; Yearley, 1991). However, what this view lacks is a sense of social, cultural and moral embeddedness of trust. Giddens's notion of trust in expert systems concerns only the rational, cognitive dimension of the social, in which 'knowing-how' is translated into a series of purely technical parameters (McKechnie and Welsh, 1993; Welsh, 2001). Indeed, what Giddens is actually talking about is not trust but reliability, the functional performativity of objects in a purely rationalized and sterile world-of-things.

Reliability relies on an instrumental sense of fidelity: a performance according to expectations. High-fidelity equipment enables the user to exercise a large amount of control and fine-tuning of the performance of the product through faultless operations of the interface. Doing the job well, which – it could be argued, is the principle of ethics in actor networks and the basis of its valorization, is thus understood in terms of an instrumental response – doing what is expected. It is the conversion of input into output, of pushing a button into obtaining results. Central to instrumental fidelity is the absence of ambivalence, ambiguity and noise.

Whilst dominant in technological culture, instrumental fidelity is certainly not the only form of trust available in modern western society. There has always been another and more transcendental sense of trust, that manifests itself in a divinely inspired sense of fidelity (*fides*). Fidelity as *fides* does not call on us to respond as expected; instead it stems from faith (*fides* also means faith). *Fides* immediately highlights the relational aspects of this modality of being. In this sense, being faithful (e.g. in marriage) is a particular response to a moral obligation that stems from one's relationship with God. Rather than an ordering/revealing of a world of things, i.e. materiality, *fides* modulates relationships and attunes them to a transcendental God. This makes *fides* antithetical to technological culture, for it reverses the relationship between human being and nature, history and the divine. These are no longer a matter of bringing forth, of challenging, revealing, regulating, securing, but become a call to which we are asked to respond (Vattimo, 1999). Fidelity as *fides* does not ask us to be autonomous, it asks us to be dependent on and open to others, it shows us that we need others as others need us.

Perhaps it is too obvious, but a transcendent apocalypse culture that directs itself against the moral degeneration of a civilization is more likely to embrace *fides* rather than instrumental fidelity. And in doing so, it invokes a transcendental God with whom we can only come into complete communion in the fullness of time, i.e. at the end of history. Hence, it is no surprise that transcendent apocalypse culture usually invokes not just the spectre of a catastrophic ending, but also that of a dawning of a new era. This notion of apocalypse as revelation can already be seen in the immanent apocalypse culture of nihilism, but there it remains forever deferred as it is preoccupied with exposure and destruction.

Technological culture, under attack from both forms of apocalypse culture, however, is not defenceless. It calls to arms its own arsenal of ballistics, including those of a moral and ethical nature, not just to defend instrumental fidelity but to export it, to use it to colonize remnants of the life world where it does not yet prevail. This is illustrated, for example, by the way in which parenting is now rapidly turning into an apparatus, a technological system, in which expertise combined with government regulation, mediation, and commercial valorization asserts itself over and above an implicit natural law which grants sovereignty to the biological parents over the upbringing of their children within limits defined by morality (Strathern, 1992). Family planning is another example of how an expert system evolves in replacing a divinely inspired sense of fidelity (e.g. life as a gift from God), into a rationally managed and controlled operation. Perhaps no example is more extreme than abortion in which a conceived life is wilfully terminated,[4] usually by rather cruel means, often with no more of an apology than its inconvenience to the mother, father or, in the case of severely handicapped babies, society at large. Crucial for the wider social acceptance of abortion, however, has been that the life that is being terminated is not defined as life, but as a mere accumulation of cells, nothing more than a cancerous growth, which has to be cut out of the woman's womb.

Genetic engineering, family planning, abortion, euthanasia, etc., are mere examples of technological culture's ongoing advance in reconstituting the meaning of life itself. The apocalypse is not coming, it is already happening. The risk society is its apparent manifestation, but this is only the beginning. Technological culture may be under attack, but it is striking back with a vengeance. Risks are not their enemies, although they are not entirely trustworthy allies. Risks are the likely battleground for the culture clash between technocracy and apocalypses. We 'humans' merely stand on the sidelines as spectators. Whereas 'man' still thinks of 'himself' as the choreographer of this epic battle, 'he' is merely a site, an arena, a vehicle, a vector, a lubricant.

Dangers and saving powers

The risk society shows us what Actor Network Theory and biophilosophy have always claimed: technological processes are not always smooth and operationally functional. Smoothness is not immanent to technological systems as their compatibility is never to be taken for granted; they have a tendency to autopoiesis as well as symbiosis. We only have to think about the combination of sabres and horses, fleas and rats, which created a great advance in military force for the nomadic Mongol army that rode across the steppes of Central Asia in the Middle Ages, in order to understand the pathological energy released by such a volatile mixture of genomes and technologies, which resulted in the Black Death (McNeill, 1976/1994) decimating the European population. More recently, of course, we have vigilante media who use helicopters, cars, hidden cameras, CCTV, etc., to connect video-recording with movement in public spaces for surveillance/entertainment purposes.

Chapter 9 discussed the relationship between a complex assemblage of police technologies, domestic entertainment technologies, litigation technologies and broadcasting technologies to describe how assemblages may result in what McLuhan (1964) refers to as hybrid media. In the aforementioned case, these media released an energy that took the form of violence in the 1992 Los Angeles 'riots'. However, it did not stop there as it affected society on a much wider scale, for example through the subsequent actions taken by civil rights organizations, political movements, justice departments, governmental agencies and even the police force itself. The extended and complex technological assemblages involved here, such as that of the . . . + *beating* + *videotape* + *broadcasting* + *court proceedings* + *jury verdict* + *news report* + . . . released a multiplicity of sensibilities at once, many of which were simply incompatible. Moreover, these effects were not limited to the USA but had a global impact, as the LA 'riots' became an icon used worldwide for race relations in the western world, as well as for the explosiveness of social injustice as a threat simmering below the smooth surface of global capitalism. In any case, such global impact would have been unthinkable without instantaneous news coverage through satellite television.

What matters here is the relationship between assemblage and violence, that is to say, the risks contained (but not completely) by the technological system that aims to control all contingencies. As McNeill (1976/1994: 57) observed, conquering armies always bring their parasites and diseases to new territories; but these remain implicit side effects of their particular strategies that may not always be beneficial to them either.

> It is worth pointing out the parallels between the microparasitism of infectious disease and the macroparasitism of military operations . . . Very early in civilised history, successful raiders became conquerors, i.e., learned how to rob agriculturists in such a way as to take from them some but not all of the harvest. By trial and error a balance could and did arise, whereby

204 Conclusion: risk and apocalypse culture

cultivators could survive such predation by producing more grain and other crops than were needed for their own maintenance.

(McNeill, 1976/1994: 57)

To understand risk in a technological culture, we must look for the elements that disrupt the homogeneity and functional smoothness of the integrated system and its functional domain.

It is the specific characteristic of the latest turning of technological culture – towards an emergent apocalypse culture – that such disruptions have become more visible. New technologies, and in particular new media hybrids, allow us to 'open up' to our being-in-the-world a revelation via shock therapy, as it were. Risks thus always have a split face. On the one hand they can be mobilized for greater systemic control and integration, on the other hand they can be mobilized to enforce radical change and innovation.

McLuhan certainly did not dismiss the importance of our 'response' to technological change.

> As long as we adopt the Narcissus attitude of regarding the extensions of our own bodies as really *out there* and really independent of us, we will meet all technological challenges with the same sort of banana-skin pirouette and collapse.

(McLuhan, 1964: 73)

In other words, the assemblage of media technologies may be seen as an opening up of an event, a possibility for intervention, a revealing whose course may be enframed as 'destined' towards a particular unfolding, but such destining is only 'determined' if accompanied by a narcissus narcosis of prosthetic numbness. The 'risk' of prosthetic numbness may thus very well be engendered by the ethos of risk aversion that inhabits large domains of our technological culture.

This understanding of media technologies consequently accomplishes two important interventions into modern thought and its rather suspect dualistic politics of philosophical fundamentalism. First, the notion of technology as ordering, revealing and enframing, radically evacuates the implicit anthropocentric understanding of being-in-the-world and its historical unfolding. Second, it breaks with the modernist tradition that reason is destined towards the *necessary* attunement of truth and justice. 'Progress' has marked not the increased autonomy of the human, but instead has been the unfolding of an increased self-enclosure of technology (*autopoiesis*), turning everything into a standing reserve, including itself. This seems to resonate a particular anti-technological romanticism which appropriates the same technological determinism as the technocrats it despises (Feenberg, 1991). Certainly, Heidegger's pessimistic writings in particular are often seen as leaning towards technological determinism and an almost mystical romanticism. However, this is not a necessary ethos in which enframing/revealing needs to be cast. McLuhan's notion of hybrid energy also reveals a sensibility towards possibilities for radical intervention – a turning against the narcissus narcosis. And, indeed, Heidegger himself wrote:

> [A]nd yet – in all the disguising belonging to Enframing, the bright-open space of the world lights up, the truth of Being flashes. And the instant, that is, when Enframing lights up, in its coming to presence, as the danger, i.e., as the saving power.
>
> (Heidegger, 1977: 47)

The question now becomes how to 'seize' the moment of this flash of light, this opening up, this release of hybrid energy. This is the essential question of technological ethics. It is the way in which we may encounter the enframing/ revealing of technology in a turning of its destining.

The concept of enframing (ordering, revealing) highlights that technology both 'produces' and 'manages' 'objects'; that is, technology is a cultivation of particular forms of sense-making. It makes phenomena visible, enables them to enter symbolic forms (become signified) and engages them in the attribution of value. Technological culture, therefore, engages in the production and management of forms of 'sense-making' that find their logic of organization not from the objects themselves, but from the assemblages in which they emerge. Technological culture is a particular social and symbolic organization of sense-making: one which *imposes* insight, meaning and value. It organizes our being-in-the-world world by creating phenomena. Against such an understanding of technological culture, one may pose another form of sensibility – one that does not rely on a challenging forth (an imposition of sense), but on an ability to respond, that is a 'learning to be affected'. This technological culture entails a responsiveness that is not based on instrumental fidelity but on *fides*. In this book, we have seen how this responsiveness is called for by the emergent 'risk society'. If we are to avoid a full-blown and catastrophic apocalypse, we have to engage with technology differently. Faith in reason is not a good starting point, reason in faith might be.

Notes

1 Introduction: technological culture and risk

1 The term 'man' is deliberately invoked to remind us of the full patriarchal connotations of this 'universalizing' category. This is necessary to unsettle the universalizing claims of modernist discourse which are hopelessly inadequate in dealing with sexual difference (Irigaray, 1987, 2000).

2 It is telling that the insurance industry still uses the category of 'Act of God' to pre-empt any liability in the face of 'natural hazards'.

3 Beck, U. (1986) *Risikogesellschaft: Auf dem Weg in eine andere Moderne*. Frankfurt am Main: Suhrkamp Verlag, translated edition (1992) *Risk Society: Towards a New Modernity*. London: Sage.

4 Lyotard's (1979) notion of performativity is rather different from Judith Butler's (1990; also see Lloyd, 1999) who used it to refer to the enactment of discursive regimes in the form of regularities and regulations, for example, in practices of identification such as 'being gay'. Although both share this emphasis on enactment, Lyotard's concept takes a more critical and historically specific turn by locating performativity within systemic configurations rather than subjects.

5 This does not mean, however, that our explorations will not engage with abstractions. Quite the opposite, theorizing everyday practices is itself necessarily the work of abstraction.

6 For example, there is obvious relevance in including complexity theory. The translation work of complexity theory into social and cultural theory, however, has hardly begun and it was thus not possible to engage in much detail with this important strand.

2 Cultivating risks: paradoxes in the work of Ulrich Beck

1 Elsewhere Beck (2000) further explores this idea of risk as becoming-real (also see Van Loon, 2000a).

2 Susan Cutter (1993) makes an interesting distinction between risk and hazard, in which she refers to risk as the measurements and calculations of probability that something (bad) might happen and to hazards as the wider social and technological formations of negative predicaments for both our social and natural worlds. Beck, however, bases his distinction between risk and hazard more on Luhmann (1995), who sees risk as the self-referential moment of a hazard, in which the latter enters into a process of communication and decision-making. A negative predicament without a point of decision-making is not a risk but a hazard (for a critique, see Fox, 1998: 671).

3 Although in his earlier work, Beck (1992: 52) refers to class and risk *positions*, his more recent contributions to theorizing risk have taken a more relational perspective (e.g. Beck, 2000: 213).

4 The text of this speech can be found on http://www.aids2000.org

5 In all fairness, this 'outrage' was primarily expressed by western, white, middle-class professionals for whom poverty may indeed be part of another world. However, the same people are also responsible for the implementation of many important 'world health' policies which cannot address poverty issues as this is always beyond their remit.

6 It can be questioned, however, how much of this medical 'knowledge' was actually enforced by public health administration, which was dominated by 'environmental' theories of infectious disease, championed by hygiene movements (Brüggemeier, 1989; Craddock, 2000; Latour, 1988; Roderick, 1997).

7 Contemporary writings in evolutionary biology (e.g. Anderson and May, 1982) suggest that it is possible for pathogens to remain strongly virulent, even if this kills their hosts, as long as they have other sufficient means of infecting other bodies than those that depend on the mobility of their hosts. Cars, aeroplanes and nursing staff provide such additional mobility.

8 The supplement is a concept derived from Derrida (1974: 270) who used it to refer to a simultaneous process of effacement and substitution that follows a particular addition. Whereas the idea of accumulation suggests that an addition further 'completes' or 'perfects' the initial condition, 'supplementation' takes this addition to be problematic because what is being added also disrupts the initial state because it also takes the place of a 'gap' (lack) which it was called upon to fill (also see Van Loon, 1996a).

9 Although one could ask whether other inequalities – such as those based on gender or ethnicity – are not more affected by risk than class, the logic of gender or ethnic inequality is more independent from that of industrial modernity (and scarcity), hence less attractive here as an example of the way in which scarcity is being displaced by risk.

10 Beck projects the development towards reflexivity mainly in historicist terms, that is, modernity itself engenders reflexivity. Lash's distinction between reflective and reflexive judgement provides an important intervention here. Rather than suggesting that early modernity was unreflexive, it suggests it was primarily reflective. Hence it did engage in self-confrontation, but only in terms of a determinate judgement based on a set of fixed, universal principles.

11 *Autopoiesis* is a term coined by two Chilean biologists, Maturana and Varela (1980), who used it to describe a process in which life takes form on the basis of the self-induced replication of cells (also see Ansell Pearson, 1997).

3 Enrolling risks in technocultural practices: notes on Actor Network Theory

1 In the so-called 'science wars' that raged throughout science and technology studies throughout much of the 1990s, e.g. the debate between Wolpert and Collins in the UK and the famous Skokal incident, a key issue raised was whether social scientists could actually comment on natural sciences, if they were not trained in them. The various battle lines drawn all evolved around the issue whether we must accept, a priori, the sovereignty of the scientific method. In the final analysis, this a priori can thus be revealed as an article of *faith*.

2 Latour's implicit Darwinism resembles that of Luhmann. Equally, as we shall learn, the admittance of fragility as central to actor networks resembles Luhmann's notion of the improbability of communication.

3 See, for example, Lilienfeld and Stolley (1994) for an overview of methods used in epidemiology to ascertain risks on the basis of demographics and clinical trials.

4 See Shields (1999: 3) for a more elaborate discussion of 'the virtual', which he distinguishes from the real, the actual and the possible.

4 Assemblages and deviations: biophilosophical reflections on risk

1 The problem is even more obvious in radical feminism. Its use of 'latent interests' has always met strong opposition from various women's groups, e.g. women of colour, lesbians and working-class women, who took issue with the sweeping generalizations behind claims made on 'their behalf' (Barrett, 1980). The issue of 'female desire' is perhaps the most poignant example of the limitations of the concept of 'interests' as it proves to be rather useless beyond the domain of cognition and instrumental rationality.

2 This is not to say that other forms of domination, e.g. those based on gender or ethnicity, are less important; they may be even more universal, but – in the world today – are simply less pervasive and all embracing. As modes of incorporation, their existence is more subject to denial. Very few would now openly justify modes of domination based on gender or ethnicity, whereas capitalism is often seen as a necessary evil because of an absence of alternatives. As a result, patriarchy and colonialism do not attain as much legitimacy and their hegemonic power relies on much more fragile translations. Ultimately, the only way in which these systems of domination have been able to persist is through economic means, i.e. in alliance with capitalism.

3 This is similar to the predominantly modern aesthetic gaze that constitutes the judgement of a work of art (see for example, Jordanova, 1989; Romanyshyn, 1989).

4 This is explicitly based on the work of Bernard Stiegler (1994).

5 A theoretical framework: risk as the critical limit of technological culture

1 The closest translation of *Gegenwärtigen* is perhaps representing.

2 As Derrida (1974) has argued, this is less so in writing in which the time taken in the formation of meaning is still reflected in its inscription as *graphe*.

3 This distinction is very similar to the one central in Deleuze and Guattari's (1994) last collaborative work *What is Philosophy?* The main difference is that Deleuze and Guattari speak of philosophy rather than religion.

4 As we shall see in the next chapter, waste – as resource – is no longer exclusively associated with negative value.

5 This is not to suggest that those risks that have been left out – such as those of terrorism, nuclear war, genetics, vaccination, nuclear power, sex, intimacy, rumours, careers, travel, substance abuse, stock markets and investments are of lesser significance, or could not equally be tied to one of the Four Riders. The sheer proliferation of risks has made it impossible to deal with most of them in depth. Mapping them would require an encyclopaedic approach that is far beyond the scope of this book.

6 Cultivating waste: excessive risks in an economy of opportunities

1 Although one could see genetic engineering as one of its actualizations, some advocates of nanotechnology, such as Drexler (1992), have insisted on the separation between the two.

2 The untimely is a concept from Nietzsche to designate the 'acting counter to our time and thereby acting on our time and, let's hope, for the benefit of a time to come' (cited in Deleuze, 1968/1994: xxi).

7 Emergent pathogen virulence: understanding epidemics in apocalypse culture

1 For more complete lists, see for example Ryan (1996: 349–53); and Morse (1993: 13–14). Also see the website of WHO's Communicable Disease Surveillance and Control Unit http://www.who.int/emc-documents.

2 Both McKechnie and Welsh (1993) and Preston (1995) refer to the Kinshasa Highway as 'the artery of HIV' in Africa, while reflecting on the relationship between motorway traffic and prostitution.

3 For a more detailed set of analyses of the relationships between economic 'development', technological innovation and epidemics, see Farvar and Milton (1973).

4 Science and fiction genres are of course always inherently connected; as genres, however, both scientists and literary writers are still at pains to make intransgressible distinctions. The first claims an authority of truth, the second an authority of originality and genius (aura).

5 A key problem with the collocation between 'safe sex' and condom use is that the latter have a 5–8 per cent failure rate (and up to 15 per cent among non-experienced users). Hence, the use of 'safe' is rather strange in this context (Ranjit *et al.*, 2001).

6 This notion is derived from Vattimo's (1988: 60–4) definition of art as a setting-into-work of truth.

8 Cyberrisks: telematic symbiosis and computer viruses

1 This is extremely common in secular doomsday scenarios, see for example Taylor (1970), and also features prominently in less extreme apocalyptic visions such as Ryan's (1996). It is also a central feature in family planning ideologies such as those of Mary Stopes.

2 This notion of a transition from molar to modular forms of power is central to Deleuze's (1992) short and unfinished excursion into 'societies of control'. Molar forms are self-contained and holistic, they operate on the basis of resemblance, they are repetitions of the same. In contrast, modular forms are distributed and folded, they operate on the principle of auto-disseminated multiplicity and are repetitions of a singular principle. An example of a molar form is the arabesque; an example of a modular form is the fractal.

3 This analysis is derived from M. Hardt's work on Deleuze (cited in Lury, 1998: 182–3).

4 It is not coincidental that apart from the ambivalence of signification both governance and mediascapes also share another ambivalent concept: representation. It is in this ambivalence that the notion of 'the public' operates (Dahlgren, 1995; Hartley, 1992; also see John Tulloch's (1993) very insightful historical analysis of the relationship between governance and information management in post-war Britain).

5 Discovery is equally a form of concealment, as it points towards the recovery of a revelation through a deployment of *logos*.

6 Multiplicities refer to entities which do not have an 'essential' origin, but exist only in their being always-already dispersed across an ensemble of differential forces and relations.

7 Similarities with so-called 'international terrorism' and its usage of the infrastructures of global finance and travel are obvious.

8 *Cyberjack*, Robert Lee / John A. Curtis (director), Everest Entertainment, 1995; *Cybertech P.D.*, David Lancaster / Rick King (director), Spectator Films, 1995.

9 Echoes of the first *Exorcist* film are never far away in the associations between innocence, childhood, bedrooms and demonic corruption.

10 An interesting comparison may be drawn with the particular 1970s genre of 'disaster films', which were generally situated in the 'present' world, and although many of them did reserve a special role for megalomanic science as destructive, the struggle

against the devastating consequences (which were conceptualized as environmental hazards, not risks) remained a largely low-tech and collectively heroic enterprise (see Liverman and Sherman, 1985).

11 Software manufacturers themselves toyed with the idea of using viruses to protect their products from illegal copying.

9   Race, riots and risk: media technologies and the engineering of moral panics

1 A good example of a non-minority engaged by the warfare state are right-wing militia in the USA. They, however, do appropriate the same discourse of marginalization to justify their transgression of the state's monopoly of legitimate violence.

2 The problem with the term 'race' is that, even when using it in inverted commas, one is always involved in its further reproduction, hence part of the very process of racialization one may want to oppose. Whereas this is indeed highly problematic, it is worthwhile reminding ourselves that this will happen as long as the term continues to have discursive significance, regardless of whether we want to acknowledge, affirm, deny, ignore or reject racialization.

3 Mythification is to turn that which is particular, historical and constructed into an appearance of universality, naturalness and given-ness. This definition is derived from Barthes's (1957/1993) famous work *Mythologies* (also see Gooding-Williams (1993a) with reference to the 1992 LA 'riots').

4 Indeed, this creates a different kind of risk, namely that of counter-surveillance. Counter-surveillance, however, is an effective strategy only if there are sufficient media outlets for the dissemination of its products.

5 This is derived from Lyotard (1988a: 140–2; 1988b: 48–71).

6 Because of this, the content of the officers' account of their own thinking is already secondary to the form in which it was cast. The reasonability of their violence is not a matter of their own discursive gyrations, but of temporal inversion which, in turn, is a trick displayed by the medium-hybrid which itself is deeply embedded in technological culture.

7 In Bourdieu's terms, symbolic violence refers to the closure of meaning, by which particular interpretations prevail at the expense of others and the arbitrariness of the sign is assigned a status of naturalness (Bourdieu and Passeron, 1977; Thompson, 1984). The acquisition of such a state of naturalness, however, is itself better understood as an effect of *symbolic power* (Bourdieu, 1982), because it is integral rather than differential. For a more in-depth discussion of the differences between violence and power, see Arendt (1970). Following Arendt, it can be argued that whereas violence is an effect of differentiation, power (*pouvoir*) is exercised as integration. Both are different from force (*puissance*), which is a potential or capacity, and thus static rather than kinetic, entrapped in a constellation of embodiments, rather than an effect of hybridization (violence) or affirmation (power).

10   Conclusion: risk and apocalypse culture

1 For a clear account of the relationship between technoscience and desire, see Welsh (2000).

2 It is still debatable, however, to what degree hyperreality is an erosion of reality. Hyperreality could also be seen as a way in which the real becomes larger, more detailed, more 'transparent', more visualized, breaking with the codes and conventions of modernist enframing (e.g. the genres of Science or Art). In this sense, Baudrillard (1983) may be right to suggest that hyperreality is more real than Reality.

3 In this respect, the only specifically modern aspect of this fatal conceit is that it evolves around a systemic rather than individual notion of sovereignty regarding the nature of representation.

4 Language is not incidental here. The term 'termination' is far more clinical and supposedly value-neutral than 'killing'. If the sanitized language of abortion was used to describe the organized process of termination of lives after birth, it would immediately conjure up accusations of Holocaust denial.

References

Adam, B. (1990) *Time and Social Theory*. Cambridge: Polity.

Adam, B. (1995) *Timewatch: The Social Analysis of Time*. Cambridge: Polity.

Adam, B. (1998) *Timescapes of Modernity: The Environment and Invisible Hazards*. London: Routledge.

Adam, B. (1999) 'Naturzeiten, Kulturzeiten und Gender – Zum Konzept "Timescape"' in Hofmeister, S. and Spitzner, M. (eds) *Zeitlandschaften: Perspektiven oeko-sozialer Zeitpolitik*. Stuttgart: Hirzel Verlag.

Adam, B. and Van Loon, J. (2000) Introduction in Adam, B., Beck, U. and Van Loon, J. (eds) *The Risk Society and Beyond: Critical Issues for Social Theory*. London: Sage.

Adam, B., Beck, U. and Van Loon, J. (2000) (eds) *The Risk Society and Beyond: Critical Issues for Social Theory*. London: Sage.

Adams, J. (1995) *Risk*. London: UCL Press.

Alexander, J. (1996) 'Critical Reflections on Reflexive Modernization'. *Theory, Culture & Society* 13(4): 133–8.

Allan, S. (1995) 'News, Truth and Postmodernity: Unravelling the Will to Facticity' in Adam, B. and Allan, S. (eds) *Theorizing Culture: Critique after Postmodernism*. London: UCL Press.

Allan, S. (2000) 'Risk and the Common Sense of Nuclearism' in Adam, B., Beck, U. and Van Loon, J. (eds) *The Risk Society and Beyond: Critical Issues for Social Theory*. London: Sage.

Allan, S., Adam, B. and Carter, C. (2000a) 'Introduction: The Media Politics of Environmental Risk' in Allan, S., Adam, B. and Carter, C. (eds) *Environmental Risks and the Media*. London: Routledge.

Allan, S., Adam, B. and Carter, C. (eds) (2000b) *Environmental Risks and the Media*. London: Routledge.

Allen, E. (1993) *L.A. Cops on Trial: The Rodney King Trial*. Columbia Home Video.

Althusser, L. (1971) *Lenin and Philosophy and Other Essays*. New York: New Left Books.

Anderson, A. (1997) *Media, Culture and the Environment*. London: UCL Press.

Anderson, M. (1983) *An Introduction to Epidemiology*, 2nd edn. London: Macmillan.

Anderson, R.M. and May, R. (1982) 'Coevolution of Hosts and Parasites'. *Parasitology* 85: 411–26.

Ankomah, B. (1998) 'Are 26 Million Africans Dying of AIDS? "The biggest lie of the century" under fire'. *New African* (December): 34–42.

Ansell Pearson, K. (1997) *Viroid Life: Perspectives on Nietzsche and the Transhuman Condition*. London: Routledge.

Arendt, H. (1970) *On Violence*. San Diego, CA: Harcourt Brace Jovanovich.

Baker, H. (1993) 'Scene . . . Not Heard' in Gooding-Williams, R. (ed.) *Reading Rodney King: Reading Urban Uprising*. London: Routledge.

Barrett, M. (1980) *Women's Oppression Today: The Marxist/Feminist Encounter*. London: Verso.

Barthes, R. (1957/1993) *Mythologies*. London: Vintage.

Barthes, R. (1977) *Image Music Text*. London: Fontana.

Basalla, G. (1988) *The Evolution of Technology*. Cambridge: Cambridge University Press.

Bataille, G. (1985) *Visions of Excess: Selected Writings, 1927–1939* (Stoekl, A. ed.). Minneapolis, MN: University of Minnesota Press.

Baudrillard, J. (1983) *Fatal Strategies: Crystal Revenge*. London: Pluto.

Baudrillard, J. (1993) *Symbolic Exchange and Death*. London: Sage.

Bauman, Z. (1990) 'Modernity and Ambivalence' in Featherstone, M. (ed.) *Global Culture: Nationalism, Globalization and Modernity*. London: Sage.

Beauvais, P. and Billette de Villemeur, T. (1996) *Maladie de Creutzfeldt–Jakob et Autres Maladies à Prion*. Paris: Médecine-Sciences Flammarion.

Beck, U. (1992) *Risk Society: Towards a New Modernity*. Trans. M. Ritter. London: Sage.

Beck, U. (1994) 'The Reinvention of Politics: Towards a Theory of Reflexive Modernization' in Beck, U., Giddens, A. and Lash, S. (eds) *Reflexive Modernization: Politics, Tradition and Aesthetics in the Modern Social Order*. Cambridge: Polity.

Beck, U. (1995) *Ecological Politics in an Age of Risk*. Cambridge: Polity.

Beck, U. (1996) 'World Risk Society as Cosmopolitan Society? Ecological Questions in a Framework of Manufactured Uncertainties'. *Theory, Culture & Society* 13(4): 1–32.

Beck, U. (1997) *The Reinvention of Politics: Rethinking Modernity in the Global Social Order*. Cambridge: Polity.

Beck, U. (1998) *Was ist Globalisierung?* Frankfurt am Main: Suhrkamp Verlag.

Beck, U. (2000) 'Risk Society Revisited: Theory, Politics and Research Programmes' in Adam, B., Beck, U. and Van Loon, J. (eds) *The Risk Society and Beyond: Critical Issues for Social Theory*. London: Sage.

Beck, U., Giddens, A. and Lash, S. (1994) *Reflexive Modernization: Politics, Tradition and Aesthetics in the Modern Social Order*. Cambridge: Polity.

Beck-Gernsheim, E. (1995) *The Social Implications of Bioengineering*. Atlantic Highlands, NJ: Humanities Press.

Beck-Gernsheim, E. (2000) 'Health and Responsibility: From Social Change to Technological Change and Vice Versa' in Adam, B., Beck, U. and Van Loon, J. (eds) *The Risk Society and Beyond: Critical Issues for Social Theory*. London: Sage.

Berg, M. and Mol, A. (eds) (1998) *Differences in Medicine: Unraveling Practices, Techniques, and Bodies*. Durham, NC: Duke University Press.

Berman, M. (1982) *All that is Solid Melts into Air*. New York: Penguin.

Bernstein, P. (1996) *Against the Gods: The Remarkable Story of Risk*. New York: John Wiley.

Berry, D. (2000) 'Trust in Media Practices: Towards Cultural Development' in Berry, D. (ed.) *Ethics and Media Culture*. Oxford: Focal Press.

Bey, H. (1991) *TAZ – The Temporary Autonomous Zone*. New York: Autonomedia.

Bijker, W. and Law, J. (eds) (1992) *Shaping Technology/Building Society: Studies in Sociotechnical Change*. Cambridge, MA: MIT Press.

Blattner, W.D. (1992) 'Existential Temporality in *Being and Time*. (Why Heidegger is not a Pragmatist)' in Dreyfus, H. and Hall, H. (eds) *Heidegger: A Critical Reader*. Oxford: Blackwell.

Bloor, M. (1995) *The Sociology of HIV Transmission*. London: Sage.

Boden, D. (2000) 'Worlds in Action: Information, Instantaneity and Global Futures Trading' in Adam, B., Beck, U. and Van Loon, J. (eds) *The Risk Society and Beyond: Critical Issues for Social Theory*. London: Sage.

Bogard, W.C. (1989) *Bhopal Tragedy: Language, Logic, and Politics in the Production of a Hazard*. Boulder, CO: Westview Press.

Bordo, S. (1993) *Unbearable Weight: Feminism, Western Culture, and the Body*. Berkeley, CA: University of California Press.

Bounan, M. (1991) *Le Temps du Sida*. Paris: Editions Allia.

Bourdieu, P. (1982) *Ce que parler veut dire: L'Economie des echanges linguistiques*. Paris: Fayard.

Bourdieu, P. and Passeron, J.C. (1977) *Reproduction in Education, Society and Culture*. London: Sage

Bouter, L.M. and van Dongen, M.C.J.M. (1991) *Epidemiologisch Onderzoek: Opzet en Interpretatie*. Houten: Bohn Stafley Van Lochum BV.

Breen, R. (1997) 'Risk, Recommodification and Stratification'. *Sociology* 31(3): 473–89.

Browning, B. (1998) *Infectious Rhythms: Metaphors of Contagion and the Spread of African Culture*. London: Routledge.

Brüggemeier, F.J. (1989) 'Medicine and Science' in Chant, C. (ed.) *Science, Technology and Everyday Life 1870–1950*. London: Routledge.

Butler, J. (1990) *Gender Trouble: Feminism and the Subversion of Identity*. London: Routledge.

Butler, J. (1993) 'Endangered/Endangering: Schematic Racism and White Paranoia' in Gooding-Williams, R. (ed.) *Reading Rodney King: Reading Urban Uprising*. London: Routledge.

Cann, A.J. (1997) *Principles of Molecular Virology*, 2nd edn. London: Academic Press.

Cannon, G. (1995) *Superbug. Nature's Revenge: Why Antibiotics Can Breed Disease*. London: Virgin.

Carmichael, A.G. (1997) 'Bubonic Plague: The Black Death' in Kiple, K.E. (ed.) *Plague, Pox & Pestilence: Disease in History*. London: Weidenfeld and Nicolson.

Castells, M. (1996) *The Rise of the Network Society*. Oxford: Blackwell.

Caufield, C. (1989) *Multiple Exposures: Chronicles of the Radiation Age*. London: Secker and Warburg.

Caygill, H. (2000) 'Liturgies of Fear: Biotechnology and Culture' in Adam, B., Beck, U. and Van Loon, J. (eds) *The Risk Society and Beyond: Critical Issues for Social Theory*. London: Sage.

Chambers, I. (1994) *Migrancy Culture Identity*. London: Routledge.

Chapman, G., Kumar, K., Fraser, C. and Gaber, I. (1997) *Environmentalism and the Mass Media: The North–South Divide*. London: Routledge.

Christie, A.B. (1987) *Infectious Diseases*, volume 1, 4th edn. London: Longman.

Clark, N. (1997a) 'Panic Ecology: Nature in the Age of Superconductivity'. *Theory, Culture & Society* 14(1): 77–96.

Clark, N. (1997b) 'Infowar/Ecodefense: Target-Rich Environmentalism from Desert Storm to Independence Day'. *Space and Culture* 2: 50–74.

Clark, N. (1998) 'Nanoplanet: Molecular Engineering in the Time of Ecological Crisis'. *Time & Society* 7(2): 353–68.

Close, W.T. (1995) *Ebola*. London: Arrow.

Cohen, S. (1987) *Folk Devils and Moral Panics: The Creation of the Mods and Rockers*. Oxford: Basil Blackwell.

Coleman, C.L. (1995) 'Science, Technology and Risk Coverage of Community Conflict'. *Media, Culture & Society* 17(1): 65–79.

Coley, N. (1989) 'From Sanitary Reform to Social Welfare' in Chant, C. (ed.) *Science, Technology and Everyday Life 1870–1950*. London: Routledge.

Colton, C.E. and Skinner, P.N. (1996) *The Road to Love Canal: Managing Industrial Waste before EPA*. Austin, TX: University of Texas Press.

Cook, R. (1987) *Outbreak*. London: Macmillan.

Cope, C.B., Fuller, W.H. and Willetts, S.L. (1983) *The Scientific Management of Hazardous Wastes*. Cambridge: Cambridge University Press.

Copjec, J. (1989) 'The Orthopsychic Subject: Film Theory and the Reception of Lacan'. *October* 49 (Summer): 53–71.

Copjec, J. (ed.) (1996) *Radical Evil*. London: Verso.

Cottle, S. (1993) *TV News, Urban Conflict and the Inner City*. Leicester: Leicester University Press.

Cottle, S. (1997) 'Ulrich Beck, "Risk Society" and the Media: A Catastrophic View?' *European Journal of Communication* 12(4): 429–56.

Craddock, S. (2000) *City of Plagues: Disease, Poverty and Deviance in San Francisco*. Minneapolis, MN: University of Minnesota Press.

Crenshaw, K. and Peller, G. (1993) 'Reel Time/Real Justice' in Gooding-Williams, R. (ed.) (1993) *Reading Rodney King: Reading Urban Uprising*. London: Routledge.

Crichton, M. (1969) *The Andromeda Strain*. New York: Centesis.

Curtis, N. (2001) *Against Autonomy: Lyotard, Judgement and Action*. Aldershot: Ashgate.

Cutter, S. (1993) *Living with Risk: The Geography of Ecological Hazards*. New York: Edward Arnold.

Dahlgren, P. (1995) *Television and the Public Sphere: Citizenship, Democracy and the Media*. London: Sage.

Davis, M. (1990) *City of Quartz: Excavating the Future in Los Angeles*. London: Verso.

Dayan, D. and Katz, E. (1992) *Media Events: The Live Broadcasting of History*. Cambridge, MA: Harvard University Press.

Debord, G. (1994) *The Society of the Spectacle*. New York: Zone Books.

de la Bruhèze, A. (1992) 'Closing the Ranks: Definition and Stabilization of Radioactive Wastes in the U.S. Atomic Energy Commission, 1945–1960' in Bijker, W. and Law, J. (eds) *Shaping Technology/Building Society: Studies in Sociotechnical Change*. Cambridge, MA: MIT Press.

De Lauretis, T. (1989) 'The Essence of the Triangle or, Taking the Risk of Essentialism Seriously: Feminist Theory in Italy, the U.S. and Britain'. *Differences* 1(3): 3–37.

Deleuze, G. (1992) 'Postscript on the Societies of Control'. *October* 59: 3–7.

Deleuze, G. (1968/1994) *Difference and Repetition*. London: Athlone.

Deleuze, G. and Guattari, F. (1983) *Anti-Oedipus: Capitalism and Schizophrenia I*. London: Athlone.

Deleuze, G. and Guattari, F. (1988) *A Thousand Plateaus: Capitalism and Schizophrenia II*. London: Athlone.

Deleuze, G. and Guattari, F. (1994) *What is Philosophy?* London: Verso.

Denzin, N. (1995) *The Cinematic Society*. London: Sage.

De Regt, A. (1984) *Arbeidersgezinnen en Beschavingsarbeid: Ontwikkelingen in Nederland 1870–1940*. Meppel: Boom.

Derrida, J. (1974) *Of Grammatology*. Baltimore, MD: Johns Hopkins University Press.

Derrida, J. (1978) *Writing and Difference*. Chicago: University of Chicago Press.

Derrida, J. (1981) *Dissemination*. Chicago: University of Chicago Press.

Derrida, J. (1982) *Margins of Philosophy*. Hemel Hempstead: Harvester Wheatsheaf.

Despret, V. (1998) 'The Body We Care For: Figures of Anthropo-Zoo-Genesis'. Paper presented at the Symposium: Theorizing Bodies in Medical Practices. Paris: Centre de Sociologie de l'Innovation, 9–11 September.

De Swaan, A. (1989) *Zorg en de Staat: Welzijn, Onderwijs en gezondheidszorg in Europa en de Verenigde Staten in de nieuwe tijd*. Amsterdam: Bert Bakker.

Dimmock, N.J., Griffiths, P.D. and Madeley, C.R. (eds) (1990) *Control of Virus Diseases*. Cambridge: Cambridge University Press.

Dobson, A. (1995) *Green Political Thought*. London: Routledge.

Doel, M. and Clarke, D. (1997) 'Transpolitical Urbanism: Suburban Anomaly and Ambient Fear'. *Space and Culture* 2: 13–36.

Donzelot, J. (1977) *La Police des familles*. Paris: Editions de Minuit.

Douglas, M. (1966/1984) *Purity and Danger*. London: Routledge.

Douglas, M. (1992) *Risk and Blame: Essays in Cultural Theory*. London: Routledge.

Douglas, M. and Wildavsky, A. (1982) *Risk and Culture: An Essay on the Selection of Technological and Environmental Dangers*. Berkeley, CA: University of California Press.

Dovey, J. (1996) 'The Revelation of Unguessed Worlds' in Dovey, J. (ed.) *Fractal Dreams: New Media in Social Context*. London: Lawrence and Wishart.

Doyal, L., Naidoo, J. and Wilton, T. (eds) (1994) *AIDS: Setting of a Feminist Agenda*. London: Taylor & Francis.

Drexler, K.E. (1992) *Engines of Creation: The Coming Era of Nanotechnology*. Oxford: Oxford University Press.

Durkheim, E. (1984) *The Division of Labour in Society*. London: Macmillan.

Eco, U. (1977) *A Theory of Semiotics*. London: Macmillan.

Evans, A.S. (1982) 'Epidemiological Concepts and Methods' in Evans, A.S. (ed.) *Viral Infections of Humans: Epidemiology and Control*, 2nd edn. New York: Plenum Medical Book Company.

Ewald, P. (1994) *Evolutions of Infectious Disease*. Oxford: Oxford University Press.

Eyles, J. and Woods, K.J. (1983) *The Social Geography of Medicine and Health*. London: Croom Helm.

Fanon, F. (1952/1986) *Black Skin White Masks*. London: Pluto.

Farvar, M.T. and Milton, J.P. (eds) (1973) *The Careless Technology: Ecology and International Development*. London: Tom Stacey.

Feenberg, A. (1991) *Critical Theory of Technology*. New York: Oxford University Press.

Fiske, J. (1994) *Media Matters, Everyday Culture and Political Change*. Minneapolis, MN: University of Minnesota Press.

Ford, B.J. (1996) *BSE The Facts: Mad Cow Disease and the Risk to Mankind*. London: Corgi.

Foucault, M. (1970) *The Order of Things: An Archaeology of the Human Sciences*. New York: Vintage.

Foucault, M. (1977a) *Discipline and Punish: The Birth of the Prison*. New York: Vintage.

Foucault, M. (1977b) *Language, Counter-Memory, Practice: Selected Essays and Interviews* (ed. Bouchard, D.F.). Oxford: Basil Blackwell.

Foucault, M. (1979) *The History of Sexuality. Volume 1: An Introduction*. New York: Vintage.

Foucault, M. (1980) *Power/Knowledge: Selected Essays and Interviews*. New York: Pantheon.

Fox, N. (1998) 'Risks, Hazards and Life Choices: Reflections on Health at Work'. *Sociology* 32(4): 665–87.

Franklin, S. (1997) *Embodied Progress: A Cultural Account of Assisted Conception.* London: Routledge.

Friedman, J. (1990) 'Being in the World: Globalization and Localization' in Featherstone, M. (ed.) *Global Culture. Nationalism, Globalization and Modernity.* London: Sage.

Fuller, J.G. (1974) *Fever! The Hunt for a New Killer Virus.* London: Hart-Davis, MacGibbon.

Game, A. (1991) *Undoing the Social: Towards a Deconstructive Sociology.* Buckingham: Open University Press.

Garrett, L. (1994) *The Coming Plague: Newly Emerging Diseases in a World out of Balance.* New York: Penguin.

Giddens, A. (1984) *The Constitution of Society.* Cambridge: Polity.

Giddens, A. (1985) *The Nation State and Violence. Volume Two of A Contemporary Critique of Historical Materialism.* Cambridge: Polity.

Giddens, A. (1990) *The Consequences of Modernity.* Cambridge: Polity.

Giddens, A. (1991) *Modernity and Self-Identity.* Cambridge: Polity.

Giddens, A. (1994) 'Living in a Post-Traditional Society' in Beck, U., Giddens, A. and Lash, S. (eds) *Reflexive Modernization: Politics, Tradition and Aesthetics in the Modern Social Order.* Cambridge: Polity.

Gilbert, G.N. and Mulkay, M. (1984) *Opening Pandora's Box: A Sociological Analysis of Scientists' Discourse.* Cambridge: Cambridge University Press.

Gilmore, R.W. (1993) 'Terror Austerity Race Gender Excess Theatre' in Gooding-Williams, R. (ed.) *Reading Rodney King: Reading Urban Uprising.* London: Routledge.

Gilroy, P. (1993) *The Black Atlantic.* London: Verso.

Goldberg, D.T. (1993) *Racist Culture: Philosophy and the Politics of Meaning.* Oxford: Basil Blackwell.

Gooding-Williams, R. (1993a) 'Look, a Negro!' Gooding-Williams, R. (ed.) *Reading Rodney King: Reading Urban Uprisings.* New York: Routledge.

Gooding-Williams, R. (ed.) (1993b) *Reading Rodney King: Reading Urban Uprising.* London: Routledge.

Gorz, A. (1982) *Farewell to the Working Class: An Essay on Post-Industrial Socialism.* London: Pluto.

Gramsci, A. (1971) *Selections from the Prison Notebooks.* New York: International Publishers.

Grist, N.R., Bell, E.J., Follett, E.A.C. and Urquhart, G.E.D. (1979) *Diagnostic Methods in Clinical Virology*, 3rd edn. Oxford: Blackwell.

Habermas, J. (1981) *Theorie des kommunikativen Handels.* Frankfurt am Main: Suhrkamp Verlag.

Habermas, J. (1989) *The Structural Transformation of the Public Sphere: An Inquiry into a Category of Bourgeois Society.* Cambridge: Polity.

Hacking, I. (1983) *Representing and Intervening: Introductory Topics in the Philosophy of Natural Science.* Cambridge: Cambridge University Press.

Hall, S., Critcher, C., Jefferson, T., Clark J. and Roberts, B. (1978) *Policing the Crisis: Mugging, the State and Law and Order.* London: Macmillan.

Hannigan, J. (1995) *Environmental Sociology: A Social Constructionist Perspective.* London: Routledge.

Hansen, A. (ed.) (1993) *The Mass Media and Environmental Issues.* Leicester: Leicester University Press.

Haour-Knipe, M. and Rector, R. (eds) (1996) *Crossing Borders: Migration, Ethnicity and AIDS*. London: Taylor and Francis.

Haraway, D. (1988) 'Situated Knowledges: The Sciences Question in Feminism and the Privilege of Partial Perspective'. *Feminist Studies* 14(3): 575–99.

Haraway, D. (1989) *Primate Visions: Gender, Race and Nature in the World of Modern Science*. New York: Routledge.

Haraway, D. (1990) 'A Manifesto for Cyborgs: Science, Technology and Socialist Feminism in the 1980s' in Nicholson, L.J. (ed.) *Feminism/Postmodernism*. New York: Routledge, Chapman and Hall.

Haraway, D. (1991) *Simians, Cyborgs and Women*. London: Free Association Books.

Haraway, D. (1997) *Modest Witness @ Second Millennium: FemaleMan© Meets Oncomouse™*. London: Routledge.

Harding, S. (1986) *The Science Question in Feminism*. Milton Keynes: Open University Press.

Hartley, J. (1992) *Teleology: Studies in Television*. London: Routledge.

Haugeland, J. (1992) 'Dasein's Disclosedness' in Dreyfuss, H. and Hall, H. (eds) *Heidegger: A Critical Reader*. Oxford: Basil Blackwell.

Hazarika, S. (1987) *Bhopal: The Lessons of a Tragedy*. New Delhi: Penguin.

Heelas, P., Lash, S. and Morris, P. (eds) (1996) *Detraditionalization: Critical Reflections on Authority and Identity*. Oxford: Blackwell.

Heidegger, M. (1927/1986) *Sein und Zeit*. Tübingen: Max Niemeyer Verlag.

Heidegger, M. (1977) *The Question Concerning Technology and Other Essays*. New York: Harper & Row.

Heidegger, M. (1954/1991) *Over Bouwen, Wonen en Denken*. Nijmegen: Sun.

Henderson, D.A. (1993) 'Surveillance Systems and Intergovernmental Cooperation' in Morse, S. (ed.) *Emerging Viruses*. New York: Oxford University Press.

Herlihy, D. (1997) *The Black Death and the Transformation of the West*. Cambridge, MA: Harvard University Press.

Hillis, K. (1996) 'A Geography of the Eye: The Technologies of Virtual Reality' in Shields, R. (ed.) *Cultures of Internet: Virtual Spaces, Real Histories, Living Bodies*. London: Sage.

Hofmeister, S. and Spitzner, M. (eds) (1999) *Zeit-Landschaften: Perspectiven öko-sozialer Zeitpolitik*. Stuttgart: Hirzel Verlag.

Holland, J. (1993) 'Replication Error, Quasispecies Populations, and Extreme Evolution Rates of RNA Viruses' in Morse, S. (ed.) *Emerging Viruses*. New York: Oxford University Press.

Holland, S. (1995) 'Descartes Goes to Hollywood: Mind, Body and Gender in Contemporary Cyborg Cinema'. *Body & Society* 1(3/4): 157–74.

Holmes, J.R. (1981) *Refuse Recycling and Recovery*. Chichester: John Wiley.

Horrox, R. (1994) *The Black Death*. Manchester: Manchester University Press.

Hunt, D.M. (1997) *Screening the Los Angeles 'Riots': Race, Seeing and Resistance*. Cambridge: Cambridge University Press.

Irigaray, L. (1987) *This Sex which is not One*. Ithaca, NY: Cornell University Press.

Irigaray, L. (2000) *To Be Two*. London: Athlone.

Ironstone-Catterall, P. (2000) 'Epidemic Logic or HIV/AIDS: The Disciplines and Crises of Knowing'. *Space and Culture* 8: 142–54.

Irwin, A. (2000) 'Risk, Technology and Modernity: Re-Positioning the Sociological Analysis of Nuclear Power' in Adam, B., Beck, U. and Van Loon, J. (eds) *The Risk Society and Beyond: Critical Issues for Social Theory*. London: Sage.

Irwin, A. and Wynne, B. (eds) (1996) *Misunderstanding Science? The Public Reconstruction of Science and Technology.* Cambridge: Cambridge University Press.

Jameson, F. (1994) *Postmodernism or the Cultural Logic of Late Capitalism.* London: Verso.

Johnson, K. (1982) 'African Hemorrhagic Fevers due to Marburg and Ebola Viruses' in Evans, A. (ed.) *Viral Infections of Humans: Epidemiology and Control,* 2nd edn. New York: Plenum Medical.

Jones, C. (1985) *Patterns of Social Policy: An Introduction to Comparative Analysis.* London: Tavistock.

Jordanova, L. (1989) *Sexual Visions: Images of Gender in Science and Medicine between the Eighteenth and Twentieth Centuries.* Hemel Hempstead: Harvester Wheatsheaf.

Kant, I. (1781/1988) *Critique of Pure Reason.* London: J.M. Dent & Sons.

Keith, M. (1991) 'Policing a Perplexed Society? No-go Areas and the Mystification of Police–Black Conflict' in Cashmore, E. and McLaughlin, E. (eds) *Out of Order? Policing Black People.* London: Routledge.

Keith, M. (1992) 'Angry Writing: (Re)presenting the Unethical World of the Ethnographer'. *Society and Space* 10: 551–68.

Keith, M. (1993) *Race, Riots and Policing: Lore and Disorder in a Multi-racist Society.* London: UCL Press.

Kelly, K. (1997) 'New Rules for the New Economy'. *Wired* 5.09 (September): http://www.wired.com/wired/5.09/newrules_pr.html

Kline, R. and Pinch, T. (1999) 'The Social Construction of Technology' in Mackenzie, D. and Wajcman, J. (eds) *The Social Shaping of Technology.* Buckingham: Open University Press.

Knorr-Cetina, K.D. (1981) *The Manufacture of Knowledge: An Essay on the Constructivist and Contextual Nature of Science.* Oxford: Pergamon.

Knorr-Cetina, K.D. (1983) 'The Ethnographic Study of Scientific Work: Towards a Constructivist Interpretation of Science' in Knorr-Cetina, K. and Mulkay, M. (eds) *Science Observed: Perspectives on the Social Study of Science.* London: Sage.

Knorr-Cetina, K. and Mulkay, M. (eds) (1983) *Science Observed: Perspectives on the Social Study of Science.* London: Sage.

Kraft, M.E. (1988) 'Analyzing Technological Risks in Federal Regulatory Agencies' in Kraft, M.E. and Vig, N.J. (eds) *Technology and Politics.* Durham, NC: Duke University Press.

Krimsky, S. (1992) 'The Role of Theory in Risk Studies' in Krimsky, S. and Golding, D. (eds) *Social Theories of Risk.* Westport, CT: Praeger.

Krimsky, S. and Golding, D. (eds) (1992) *Social Theories of Risk.* Westport, CT: Praeger.

Kristeva, J. (1988) *Etrangers à nous-mêmes.* Paris: Fayard.

Kroker, A. and Cook, D. (1988) *The Postmodern Scene: Excremental Culture and Hyperaesthetics.* London: Macmillan.

Kuhn, Th. S. (1970) *The Structure of Scientific Revolutions.* Chicago: University of Chicago Press.

Lacan, J. (1977) *Ecrits: A Selection.* London: Routledge.

Lakatos, I. (1978) *The Methodology of Scientific Research Programmes.* Cambridge: Cambridge University Press.

Land, N. (1995) 'Machines and Technocultural Complexity: The Challenge of the Deleuze–Guattari Conjunction'. *Theory, Culture & Society* 12(2): 131–40.

Lash, S. (1990) *The Sociology of Postmodernism.* London: Routledge.

Lash, S. (1994) 'Reflexivity and its Doubles: Structure, Aesthetics, Community' in Beck, U., Giddens, A. and Lash, S. (eds) *Reflexive Modernization: Politics, Tradition and Aesthetics in the Modern Social Order*. Cambridge: Polity.

Lash, S. (2000) 'Risk Culture' in Adam, B., Beck, U. and Van Loon, J. (eds) *The Risk Society and Beyond: Critical Issues for Social Theory*. London: Sage.

Lash, S. and Urry, J. (1994) *Economies of Signs and Space*. London: Sage.

Latour, B. (1987) *Science in Action: How to Follow Scientists and Engineers Through Society*. Milton Keynes: Open University Press.

Latour, B. (1988) *The Pasteurization of France*. Cambridge, MA: Harvard University Press.

Latour, B. (1993) *We Have Never Been Modern*. Hemel Hempstead: Harvester Wheatsheaf.

Latour, B. (1998) 'How to be Iconophilic in Art, Science and Religion?' in Jones, C.A. and Galison, P. with A. Slaton (eds) *Picturing Science Producing Art*. New York: Routledge.

Latour, B. (1999a) '"Thou shalt not take the Lord's name in vain" – Being a Sort of Sermon in the Sociology of Religion'. Paper presented at the Sociality/Materiality Conference, Brunel University, London, 9–11 September 1999.

Latour, B. (1999b) 'Why Angels Do Not Make Good Scientific Instruments'. Unpublished paper.

Latour, B. and Woolgar, S. (1979) *Laboratory Life: The Social Construction of Scientific Facts*. London: Sage.

Law, J. (1995) 'Organization and Semiotics: Technology, Agency and Representation' in Mouritsen, J. and Munro, R. (eds) *Accountability, Power and Ethos*. London: Chapman and Hall.

Lederberg, J. (1993) 'Viruses and Humankind: Intracellular Symbiosis and Evolutionary Conception' in Morse, S. (ed.) *Emerging Viruses*. Oxford: Oxford University Press.

Levine, A. (1992) *Viruses*. New York: Scientific American Library.

Levitas, R. (2000) 'Discourses of Risk and Utopia' in Adam, B., Beck, U. and Van Loon, J. (eds) *The Risk Society and Beyond: Critical Issues for Social Theory*. London: Sage.

Lewis, J. (1982) 'The Story of a Riot: The Television Coverage of Civil Unrest in 1981'. *Screen Education* 40 (autumn/winter): 15–33.

Lilienfeld, D.A. and Stolley, P.D. (1994) *Foundations of Epidemiology*, 3rd edn. Oxford: Oxford University Press.

Lipsitz, G. (1994) 'We Know What Time It Is: Race, Class and Youth Culture in the Nineties' in Ross, A. and Rose, T. (eds) *Microphone Fiends: Youth Music and Youth Culture*. London: Routledge.

Liverman, D.M. and Sherman, D.J. (1985) 'Natural Hazards in Novels and Films: Implications for Hazard Perception and Behavior' in Burgess, J. and Gold, J.R. (eds) *Geography, the Media and Popular Culture*. Beckenham: Croom Helm.

Lloyd, M. (1999) 'Performativity, Parody, Politics'. *Theory, Culture & Society* 16(2): 195–213.

Lodziak, C. (1995) *Manipulating Needs: Capitalism and Culture*. London: Pluto.

Luhmann, N. (1982) *The Differentiation of Society*. New York: Columbia University Press.

Luhmann, N. (1990) *Essays on Self-Reference*. New York: Columbia University Press.

Luhmann, N. (1995) *Die Soziologie des Risikos*. Berlin: De Gruyter.

Lundell, A. (1989) *Virus*. Chicago: Contemporary Books.

Lupton, D. (1994) *Medicine as Culture: Illness, Disease and the Body in Western Societies*. London: Sage.

Lupton, D. (1999) *Risk*. London: Routledge.

Lury, C. (1998) *Prosthetic Culture: Photography, Memory and Identity*. London: Routledge.

Lyotard, J-F. (1971) *Discourse/Figure*. Paris: Klincksiek.

Lyotard, J-F. (1979) *La Condition postmoderne: Rapport sur le savoir*. Paris: Editions de Minuit.

Lyotard, J-F. (1988a) *The Differend: Phrases in Dispute*. Manchester: Manchester University Press.

Lyotard, J-F. (1988b) *Le Postmoderne expliqué aux enfants: Correspondence 1982–1985*. Paris: Edition Galilée.

Lyotard, J-F. (1991) *The Inhuman: Reflections on Time*. Trans. G. Bennington and R. Bowlby. Cambridge: Polity.

McCarthy, C. (1997) 'Constructions of a Culinary Abject'. *Space and Culture* 1 (July): 9–23.

McCormick, J.B. and Fisher-Hoch, S. with Horvitz, L.A. (1996) *Level 4: Virus Hunters of the CDC*. Atlanta, GA: Turner.

McKechnie, R. and Welsh, I. (1993) 'Between the Devil and the Deep Green Sea: Defining Risk Societies and Global Threats' in Weeks, J. (ed.) *The Lesser Evil, The Greater Good*. London: Rivers Oram.

Mackenzie, D. and Wajcman, J. (eds) (1999) *The Social Shaping of Technology*, 2nd edn. Buckingham: Open University Press.

McLuhan, M. (1964) *Understanding Media. The Extensions of Man*. Harmondsworth: Penguin.

McLuhan, M. and Fiore, Q. (1967) *The Medium is the Massage*. Harmondsworth: Penguin.

McMurray, C. and Smith, R. (2001) *Diseases of Globalization: Socioeconomic Transitions and Health*. London: Earthscan.

Macnaghten, Ph. and Urry, J. (1998) *Contested Natures*. London: Sage.

McNeill, W.H. (1976/1994) *Plagues and Peoples*. Harmondsworth: Penguin.

Maffesoli, M. (1996) *The Time of the Tribes: The Decline of Individualism in Mass Society*. London: Sage.

Managers Ontmoeting Overheid Bedrijfsleven (1993) *Afvalmanagement en Afvalpolitiek: Hoe krijgen we de Afvalberg binnen de Perken?* The Hague: SMO.

Mann, J., Tarantola, D.J.M. and Netter, Th. W. (eds) (1992) *AIDS in the World*. Cambridge, MA: Harvard University Press.

Mann, J., Gruskin, A.S., Grodin, M. and Annas, G.J. (eds) (1999) *Health and Human Rights: A Reader*. London: Routledge.

Margulis, L. (1993) *Symbiosis in Cell Evolution: Microbial Communities in the Archean and Proterozoic Eons*, 2nd edn. New York: Freeman.

Martin, E. (1987) *The Woman in the Body: A Cultural Analysis of Reproduction*. Milton Keynes: Open University Press.

Marx, K. (1984) *Het Kapitaal: Een Kritische Beschouwing over de Economie. Deel I: Het Reproductieproces van het Kapitaal*. Weesp: De Haan.

Marx, K. and Engels, F. (1988) *Het Communistisch Manifest*. Amsterdam: Pegasus.

Marx, W. (1971) *Man and his Environment: Waste*. New York: Harper & Row.

Maturana, H. and Varela, F. (1980) *Autopoiesis and Cognition: The Realization of the Living*. London: Reidel.

Mauss, M. (1990) *The Gift: The Form and Reason for Exchange in Archaic Societies*. London: Routledge.

Meadows, D.H., Meadows, D., Randers, J. and Behrens, W. (1972) *Limits to Growth*. London: Earth Island.

Mellanby, K. (1972) *The Biology of Pollution*. London: Edward Arnold.

Miles, R. (1989) *Racism*. London: Routledge.

Mol, A. (1998) 'Missing Links, Making Links: The Performance of some Atheroscleroses' in Berg, M. and Mol, A. (eds) *Differences in Medicine: Unraveling Practices, Techniques, and Bodies*. Durham, NC: Duke University Press.

Mormont, M. and Dasnoy, C. (1995) 'Source strategies and the Mediatization of Climate Change'. *Media, Culture & Society* 17(1): 49–64.

Morse, S.S. (ed.) (1993) *Emerging Viruses*. Oxford: Oxford University Press.

Mulkay, M. (1979) *Science and the Sociology of Knowledge*. London: George Allen & Unwin.

Negroponte, N. (1995) *Being Digital*. New York: Knopf.

Nietzsche, F. (1983) *Waarheid en Cultuur: Een keuze uit het vroege werk*. Boom: Meppel.

Nietzsche, F. (1989) *Beyond Good & Evil: Prelude to a Philosophy of the Future*. New York: Vintage.

Nietzsche, F. (1990) *Twilight of the Idols/The Anti-Christ*. Harmondsworth: Penguin.

Nowotny, H. (1994) *Time: the Modern and Postmodern Experience*. Cambridge: Polity.

Oppenheimer, J. (1997) 'Movements, Markets and the Mainstream: Gay Activism and Assimilation in the Age of AIDS' in Oppenheimer, J. and Rickitt, H. (eds) *Acting on AIDS: Sex, Drugs and Politics*. London: Serpent's Tail.

Oppenheimer, J. and Rickitt, H. (eds) (1987) *Acting on AIDS: Sex, Drugs and Politics*. London: Serpent's Tail.

Pacey, A. (1983) *The Culture of Technology*. Oxford: Basil Blackwell.

Pagels, E. (1995) *The Origins of Satan*. New York: Vintage.

Paillard, B. (1998) *Notes on the Plague Years: AIDS in Marseilles*. New York: Aldine de Gruyter.

Parfray, A. (1990) *Apocalypse Culture*. Los Angeles: Feral House.

Parsons, T. (1973) *The Evolution of Societies*. Englewood Cliffs, NJ: Prentice Hall.

Pastore, J.L. (ed.) (1993) *Confronting AIDS through Literature: The Responsibilities of Representation*. Urbana, IL: University of Illinois Press.

Patton, C. (1990) *Inventing AIDS*. London: Routledge.

Pêcheux, M. (1982) *Language, Semantics and Ideology: Stating the Obvious*. London: Macmillan.

Peters, C.J., Johnson, E.D., Jahrling, P.B., Ksiazek, T.G., Rollin, P.E., White, J., Hall, W., Trotter, R. and Jaax, N. (1993) 'Filoviruses' in S. Morse (ed.) *Emerging Viruses*. New York: Oxford University Press.

Phillips, N.A. (2000) 'Report of the BSE Inquiry' http://www.bseinquiry.gov.uk (accessed on 30 November 2001).

Poster, M. (1995) 'Postmodern Virtualities'. *Body & Society* (3/4): 79–95.

Preston, R. (1995) *The Hot Zone*. London: Transworld.

Prior, L. (2000) 'Mathematics, Risk and Genetics' in Adam, B., Beck, U. and Van Loon, J. (eds) *The Risk Society and Beyond: Critical Issues for Social Theory*. London: Sage.

Ranjit, N., Bankole, A., Darroch, J.E. and Singh, S. (2001) 'Contraceptive Failure in the First Two Years of Use: Differences across Socioeconomic Subgroups'. *Family Planning Perspectives* 33(1): 19–27.

Rasnick, D. (2000) 'A World of Fraud: Science and Medicine Out of Africa (Interview)'. *San Francisco Herald* (October) http://home.no.net/boreg/Hoax/rasnick2.htm (accessed on 30 November 2001).

Rattansi, A. (1994) '"Western" Racisms, Ethnicities and Identities in a "Postmodern" Frame' in Rattansi, A. and Westwood, S. (eds) *Racism, Modernity and Identity on the Western Front*. Cambridge: Polity.

Rayner, S. (1992) 'Cultural Theory and Risk Analysis' in Krimsky, S. and Golding, D. (eds) *Social Theories of Risk*. Westport, CT: Praeger.

Rheingold, H. (1991) *Virtual Reality*. New York: Simon & Schuster.

Rheingold, H. (1994) *The Virtual Community: Finding Connection in a Computerised World*. London: Secker and Warburg.

Rhodes, L.A. (1998) 'Panoptic Intimacies'. *Public Culture* 10(2): 285–311.

Ritenour, E.R. (1984) 'Overview of the Hazards of Low-Level Exposure to Radiation' in Hendee, W.R. (ed.) *Health Effects of Low Level Radiation*. Norwalk, CT: Appleton-Century-Crofts.

Robins, K. (1996) 'Cyberspace and the World We Live in' in Dovey, J. (ed.) *Fractal Dreams: New Media in Social Context*. London: Lawrence and Wishart.

Roderick, I. (1997) 'Household Sanitation and the Flow of Domestic Space'. *Space and Culture* 1 (Flow): 105–32.

Rogozinski, J. (1996) 'It Makes us Wrong: Kant and Radical Evil' in Copjec, J. (ed.) *Radical Evil*. London: Verso.

Romanyshyn, R.D. (1989) *Technology as Symptom and Dream*. London: Routledge.

Rose, H. (2000) 'Risk, Trust and Scepticism in the Age of New Genetics' in Adam, B., Beck, U. and Van Loon, J. (eds) *The Risk Society and Beyond: Critical Issues for Social Theory*. London: Sage.

Ryan, F. (1996) *Virus X: Understanding the Real Threat of Pandemic Plagues*. London: HarperCollins.

Salyers, A. and Whitt, D.D. (1994) *Bacterial Pathogenesis: A Molecular Approach*. Washington, DC: ASM Press.

Sapp, J. (1994) *Evolution by Association: A History of Symbiosis*. Oxford: Oxford University Press.

Sayer, D. (1991) *Capitalism and Modernity: An Excursus on Marx and Weber*. London: Routledge.

Schneider, M. and Geißler, K.A. (eds) (1999) *Flimmerende Zeiten: Vom Tempo der Medien*. Stuttgart: Hirzel Verlag.

Schwaller, C. (1997) 'Year 2000. A Date with Destiny. Apocalypse as "The End" or as "Revelation"?' *Space and Culture* 2 (Apocalypse): 37–49.

Scott, A. (2000) 'Risk Society or Angst Society: Two Views of Risk, Consciousness and Community' in Adam, B., Beck, U. and Van Loon, J. (eds) *The Risk Society and Beyond: Critical Issues for Social Theory*. London: Sage.

Shannon, S. and Weaver, W. (1949) *The Mathematical Theory of Communication*. Urbana, IL: University of Illinois Press.

Shields, R. (1997) 'Flow'. *Space and Culture* 1 (Flow as a New Paradigm): 1–7.

Shields, R. (1999) 'Virtual Spaces?' *Space and Culture* 4/5 (Virtual Space and Organizational Networks): 1–12.

Shilts, R. (1987) *And the Band Played On: Politics, People and the AIDS Epidemic*. New York: Penguin.

Shope, R.E. and Evans, A.S. (1993) 'Assessing Geographic and Transport Factors, and Recognition of New Viruses' in Morse, S. (ed.) *Emerging Viruses*. New York: Oxford University Press.

Simone, A. (2000) 'Africities: Popular Engagements of the Urban in Contemporary Africa'. *Space and Culture* 9 (Dialogues): 244–56.

Slee-Smith, P.I. (1976) *Recycling Waste*. Broseley: Scientific Publications.

Smith, D. (1974) 'Women's Perspective as a Radical Critique of Sociology'. *Sociological Inquiry* 44(1): 7–13.

Smith-Hughes, S. (1977) *The Virus: A History of the Concept*. New York: Heinemann.

Solomos, J. (1988) *Black Youth, Racism and the State: The Politics of Ideology and Policy*. Cambridge: Cambridge University Press.

Sontag, S. (1989) *AIDS and its Metaphors*. London: Allen Lane.

Stevenson, N. (1995) *Understanding Media Cultures*. London: Sage.

Stiegler, B. (1994) *La Technique et le temps 1: La Faute d'épiméthée*. Paris: Galilée.

Strathern, M. (1992) *After Nature: English Kinship in the Late Twentieth Century*. Cambride: Cambridge University Press.

Swenson, J.D. (1995) 'Rodney King, Reginald Denny, and TV News: Cultural (Re-) Constructions of Racism'. *Journal of Communication Inquiry* 19(1): 75–88.

Taylor, G.R. (1970) *The Doomsday Book*. London: Thames and Hudson.

Teubner, G. (1989) *Recht als autopoietisches System*. Frankfurt am Main: Suhrkamp Verlag.

Thompson, J. (1984) *Studies in the Theory of Ideology*. Cambridge: Polity.

Thompson, J. (1990) *Ideology and Modern Culture: Critical Social Theory in the Era of Mass Communication*. Cambridge: Polity.

Thompson, J. (1995) *The Media and Modernity: A Social Theory of the Media*. Cambridge: Polity.

Thompson, K. (1998) *Moral Panics*. London: Routledge.

Tomas, D. (1995) 'Feedback and Cybernetics. Reimaging the Body in the Age of Cybernetics'. *Body & Society* 1(3/4): 21–43.

Tönnies, F. (1887/1957) *Community and Society*. New York: Harper and Row.

Toolis, K. (2001) 'Epidemic in Waiting'. *Guardian Weekend* (22 September): 20–7.

Touraine, A. (1971) *The Post-Industrial Society*. New York: Random House.

Treichler, P. (1988) 'AIDS, Homophobia and Biomedical Discourse: An Epidemic of Signification, in Crimp, D. (ed.) *AIDS: Cultural Analysis, Cultural Activism*. Cambridge, MA: MIT Press.

Trew, T. (1979) 'Theory and Ideology at Work' in Fowler, R., Hodge, R., Kress, G. and Trew, T. (eds) *Language and Control*. London: Routledge and Kegan Paul.

Tulloch, J. (1993) 'Policing the Public Sphere: the British Machinery of News Management'. *Media Culture & Society* 15(3): 363–84.

Ungar, S. (1998) 'Hot Crises and Media Reassurance: A Comparison of Emerging Diseases and Ebola Zaire'. *British Journal of Sociology* 49(1): 36–56.

Van Dijk, T.A. (1991) *Racism and the Press: Critical Studies in Racism and Migration*. London: Routledge.

Van Loon, J. (1996a) 'A Cultural Exploration of Time: Some Implications of Temporality and Mediation'. *Time & Society* 5(1): 61–84.

Van Loon, J. (1996b) 'Technological Sensibilities and the Cyberpolitics of Gender: Donna Haraway's Postmodern Feminism'. *Innovation* 9(2): 231–43.

Van Loon, J. (1996c) 'Violence and Mediation: Figuring out the Racial Matrix of the 1992 L.A. Riots'. PhD thesis, Lancaster: Lancaster University.

Van Loon, J. (1997a) 'The End of Antibiotics: Notes towards an Investigation'. *Space and Culture* 2 (Apocalypse): 127–48.

Van Loon, J. (1997b) 'Chronotopes of/in the Televisualization of the 1992 L.A. Riots'. *Theory, Culture and Society* 14(2): 89–104.

Van Loon, J. (2000a) 'Virtual Risks in an Age of Cybernetic Reproduction' in Adam, B., Beck, U. and Van Loon, J. (eds) *The Risk Society and Beyond: Critical Issues for Social Theory*. London: Sage.

Van Loon, J. (2000b) 'Mediating the Risks of Virtual Environments' in Allan, S., Adam, B. and Carter, C. (eds) *Environmental Risks and the Media*. London: Routledge.

Van Loon, J. (2002) 'A Contagious Living Fluid'. *Theory, Culture & Society*, forthcoming.

Van Loon, J. and Sabelis, I. (1997) 'Recycling Time: The Temporal Complexity of Waste Management'. *Time & Society* 6(2/3): 287–306.

Vattimo, G. (1988) *The End of Modernity: Nihilism and Hermeneutics in Postmodern Culture*. Cambridge: Polity.

Vattimo, G. (1992) *The Transparent Society*. Cambridge: Polity.

Vattimo, G. (1997) *Beyond Interpretation: The Meaning of Hermeneutics for Philosophy*. Cambridge: Polity.

Vattimo, G. (1999) *Belief*. Cambridge: Polity.

Vig, N.J. (1988) 'Technology, Philosophy, and the State: An Overview' in Kraft, M.E. and Vig, N.J. (eds) *Technology and Politics*. Durham, NC: Duke University Press.

Virilio, P. (1986) *Speed and Politics: An Essay on Dromology*. New York: Semiotext(e).

Waldby, C. (1996) *AIDS and the Body Politic: Biomedicine and Sexual Difference*. London: Routledge.

Waterson, A.P. and Wilkinson, L. (1978) *An Introduction to the History of Virology*. Cambridge: Cambridge University Press.

Weber, M. (1956) *Wirtschaft und Gesellschaft*. Tübingen: Mohr.

Weber, M. (1985) *The Protestant Ethic and the Spirit of Capitalism*. London: Unwin.

Webster, A. (1991) *Science, Technology and Society: New Directions*. London: Macmillan.

Webster, F. (1995) *Theories of the Information Society*. London: Routledge.

Welford, R. (1994) 'Pushing Forward the Frontiers: Environmental Strategies at the Body Shop International' in Welford, R. (ed.) *Cases in Environmental Management and Business Strategy*. London: Pitman.

Welford, R. and Gouldson, A. (1993) *Environmental Management and Business Strategy*. London: Pitman.

Welsh, I. (2000) 'Desiring Risk: Nuclear Myths and the Social Selection of Risk' in Adam, B., Beck, U. and Van Loon, J. (eds) *The Risk Society and Beyond: Critical Issues for Social Theory*. London: Sage.

Welsh, I. (2001) *Mobilizing Modernity: The Nuclear Moment*. London: Routledge.

Wieviorka, M. (1995) *The Arena of Racism*. London: Sage.

Wildey, P. (1987) 'Jenner, Genes, Vaccines and Black Boxes' in Russel, W.C. and Almond, J.W. (eds) *Molecular Basis of Virus Diseases*. Cambridge: Cambridge University Press.

Williams, P.J. (1993) 'The Rules of the Game' in Gooding-Williams, R. (ed.) *Reading Rodney King: Reading Urban Uprising*. London: Routledge.

Wills, C. (1996) *Yellow Fever Black Goddess: The Coevolution of People and Plagues*. New York: Addison-Wesley.

Winner, L. (1977) *Autonomous Technology: Technics-out-of-Control as a Theme in Political Thought*. Cambridge, MA: MIT Press.

World Health Organization (1996) *The World Health Report 1996: Fighting Disease, Fostering Development*. Report of the Director General. Geneva: World Health Organization.

World Health Organization (1998) *Factsheet* no. 97 (August). Geneva: World Health Organization.

Wynne, B. (1989) 'Sheepfarming after Chernobyl: A Case-Study in Communicating Scientific Information'. *Environment* 31(2): 10–15, 33–9.

Wynne, B. (1996) 'May the Sheep Safely Graze?' in Lash, S., Szerszynsky, B. and Wynne, B. (eds) *Risk, Environment and Modernity*. London: Sage.

Yearley, S. (1991) *The Green Case: A Sociology of Environmental Issues, Arguments and Politics*. London: HarperCollins.

Yearley, S. (1996) *Sociology, Environmentalism, Globalization: Reinventing the Globe*. London: Sage.

Young, R. (1994) 'Egypt in America: *Black Athena*, Racism and Colonial Discourse' in Rattansi, A. and Westwood, S. (eds) *Racism, Modernity and Identity on the Western Front*. Cambridge: Polity.

Zolo, D. (1991) 'Autopoiesis: Critique of a Postmodern Paradigm'. *Telos* 86: 61–80.

Index

body 73, 143; and infection 144; and
   subjectivity 153; and technology 69–70
Body Shop 116–17
boundaries 36, 193; between possible and
   real 154–5; immune system and 72; and
   racialization 174; telematic symbiosis
   and 158; transgressions 82
BSE (bovine spongiform encephalopathy)
   4, 58–60, 132

capital 74, 97
capitalism 9, 20, 22, 64, 65, 96;
   incorporation 67–8, 208; and *nomos* 83;
   and waste 115, 120–1
Castells, Manuel 48
chemicals 106–7
CJD 59; vCJD 58–60
Clark, Nigel 118–19
class 20–1, 26, 100–1; class struggle 8, 31,
   64
classification *see* ordering
cleanliness 28–9
closure 38, 40, 50, 97, 113
collective violence 169, 170; and the
   media 171–2; *see also* violence
collectives 64, 68
colonialism 65
command, control, communication
   intelligence (C3I) systems 167
commerce 42, 83, 114
commodification 39, 67, 115
communication 35, 36; Actor Network
   Theory and 51; and autopoiesis 36–8, 43
communicative action theory 19, 39
complexity 25–6, 35, 41, 42
complexity theory 14, 194, 206
computer viruses 158, 159, 161–6
concealment 61; and unconcealment 91,
   94, 95–6, 155, 198
condoms 24, 209
conscience collectif 20
conscious consumption 30
consciousness 10–11; collective 43–4
conservation 64, 110
constructionism, social 6, 20, 63
containment 25, 31, 45, 74, 86
contamination 123
contingency 86, 98, 113, 187, 189;
   calculability 194

control 63, 193
conversion *see* translation
credibility 97
credit, cycles of 74, 97, 113
Crichton, Michael 138; *The Andromeda
   Strain* 134–5, 138–40, 141–2
crisis management 133
cultivation 8
cultural determinism versus technological
   determinism 8–9
culture 7, 8, 193; and technology 8–11;
   western 10; *see also* apocalypse cultures
Cutter, Susan 6–7, 206
cyberfeminism 70
Cyberjack (film) 163–4, 209
cybernetics 151–2, 196
cyberrisks 158, 159, 160–1, 196; and
   analogue media 171–2; viral terrorism
   161–6
cyberspace (space of flows) 148, 150, 153
Cybertech P.D. (film) 164, 209
cyborgs: films about 165; Haraway's 69–70,
   77, 80, 167

de Bary, Anton 77
decadence 199
decisions 32; making 12, 30, 41
Deleuze, G. 72, 76, 78–9, 80, 81, 89
democracy 39, 40, 43, 44, 157
democratic despotism 179
demons 164–5
desire 186, 210
deterritorialization 78–9, 82, 154, 193
detraditionalization 26, 32, 200; *see also*
   individualization
differance 57
difference 49, 57, 80–1
differend 85
differentiation 35, 36
diffraction 34, 75, 76, 77
digitalization 150, 152
discipline 68; AIDS and 141
disease 123, 124, 125; Ebola 126–9;
   and industrial sites 29; spread of 23,
   131–2; waste and 106, 108–9; *see also*
   pathogens
divinity 92–3
Douglas, M. 7, 108
Durkheim, E. 20, 33

hybrid energy 204–5
hygiene movement 28–9, 109, 143
hyperreality 150, 195, 196, 210
hyperthought 150–1
hypocrisy 199

ICTs (information and communication
    technologies) 11–13, 167–8, 195–6;
    cyberrisks 160–6; environment of
    150–1; telematics 152–5; virtual risks
    and telematic symbiosis 156–8
identity 27, 49; immune system and 72, 75,
    77; institutional 33
immanent apocalypse culture 198–9
immune system 71–3, 77, 122
immunity 71
immunization 74, 77, 86, 191
immutable mobiles 51–3, 54, 55, 56, 60;
    waste as 114
incorporation 67–8, 69–70, 198, 208
indexical violence 181–2, 182
individualization 26–8, 29–34, 42, 188
individuation 26
induction 67–8
industrial modernity 14, 19–22, 29–31,
    188; *see also* modernity
infection 106, 144, 186, 207; *see also*
    disease; pathogens
information 37; cybernetics and 151; vast
    amount about risks 188
information and communication
    technologies *see* ICTs
informational landscape 148–9, 150–1
inhabitation 8
inhuman 84–5, 86
innovation *see* technological innovation
institutions 31, 32, 33, 39, 64; *logos* and
    nomos and 82–3; new 200–1
instrumentalism 9, 194–5
insurance 197
integration 35; lack of 173, 174–5
integrity 74, 75, 192; judgement and 179
interests 64, 65–6
intuition 178, 179, 184
invisibility 43
irreducibility 48, 49, 76, 192

judgement 178–9, 180
junctural zones 131, 147, 192

Jurassic Park (film) 138
justice 180, 201

King, Rodney, beating of and subsequent
    trial 172, 176–8
kitchens and waste 109
knowledge 41, 91–2, 98; lack of 113;
    overcoding of 157; partial 197; as
    realization 48; and vision 69
Krimsky, S. and Golding, D. 6

lack 22; of knowledge 113; and racial
    difference 175; and urban disorders 173,
    174; *see also* scarcity
Lash, Scot 34, 207
Lassa fever 128
Latour, Bruno 14, 45, 61–2, 88–9, 189,
    191; *see also* Actor Network Theory
law 38, 41, 81; law and order discourse
    173, 175
Law, John 54, 57
legal discourse 38, 180–1
liberalism 113–14
litter 110–11
logic 9–10, 156–7; communicative 51; of
    equivalence 175; of life 150
logorationality 177–8, 194
logos 81–2, 86, 156, 159–60; and
    insitutions 82–3, 84
Los Angeles, 1992 'riots' 172, 173, 203;
    beating of Rodney King and ensuing
    trial 176, 177–8
Luhmann, Niklas 19, 35–6, 37–8
Lundell, Allan 161–2, 165–6
Lyotard, Jean-François 84–6, 89, 150, 206

machinic assemblage *see* assemblage
McLuhan, M. 12
'man' 5, 153, 206; losing faith in 33;
    modern 2–3, 10, 113, 174; and nature 9,
    149, 198
mapping 50–1
Marburg virus 128
Margulis, Lynn 77, 79
Marxism 20, 21, 26, 64
mass media 11–12, 40, 171, 200; *see also*
    media
matters of fact 49, 71
Mbeki, President 22